The Open University

Science Foundation Course Unit 17

**THE GENETIC CODE:
GROWTH AND REPLICATION**

Prepared by the Science Foundation Course Team

THE OPEN UNIVERSITY PRESS

A NOTE ABOUT AUTHORSHIP OF THIS TEXT

This text is one of a series that, together, constitutes *a component part* of the Science Foundation Course. The other components are a series of television and radio programmes, home experiments and a summer school.

The course has been produced by a team, which accepts responsibility for the course as a whole and for each of its components.

THE SCIENCE FOUNDATION COURSE TEAM

M. J. Pentz (Chairman and General Editor)

F. Aprahamian	(Editor)	A. R. Kaye	(Educational Technology)
A. Clow	(BBC)	J. McCloy	(BBC)
P. A. Crompton	(BBC)	J. E. Pearson	(Editor)
G. F. Elliott	(Physics)	S. P. R. Rose	(Biology)
G. C. Fletcher	(Physics)	R. A. Ross	(Chemistry)
I. G. Gass	(Earth Sciences)	P. J. Smith	(Earth Sciences)
L. J. Haynes	(Chemistry)	F. R. Stannard	(Physics)
R. R. Hill	(Chemistry)	J. Stevenson	(BBC)
R. M. Holmes	(Biology)	N. A. Taylor	(BBC)
S. W. Hurry	(Biology)	M. E. Varley	(Biology)
D. A. Johnson	(Chemistry)	A. J. Walton	(Physics)
A. B. Jolly	(BBC)	B. G. Whatley	(BBC)
R. Jones	(BBC)	R. C. L. Wilson	(Earth Sciences)

The following people acted as consultants for certain components of the course:

J. D. Beckman	B. S. Cox
H. G. Davies	R. J. Knight
D. J. Miller	M. W. Neil
J. R. Ravetz	H. Rose

The Open University Press,
Walton Hall, Bletchley, Buckinghamshire

First Published 1971
Copyright © 1971 The Open University

Designed by The Media Development Group of the Open University

Printed in Great Britain by
Staples Printing Group
at the Priory Press, St Albans, Herts.

SBN 335 02008 9

Open University courses provide a method of study for independent learners through an integrated teaching system, including textual material, radio and television programmes and short residential courses. This text is one of a series that make up the correspondence element of the Science Foundation Course.

The Open University's courses represent a new system of university level education. Much of the teaching material is still in a developmental stage. Courses and course materials are, therefore, kept continually under revision. It is intended to issue regular up-dating notes as and when the need arises, and new editions will be brought out when necessary.

Further information on Open University courses may be obtained from The Admissions Office, The Open University, P.O. Box 48, Bletchley, Buckinghamshire.

Contents

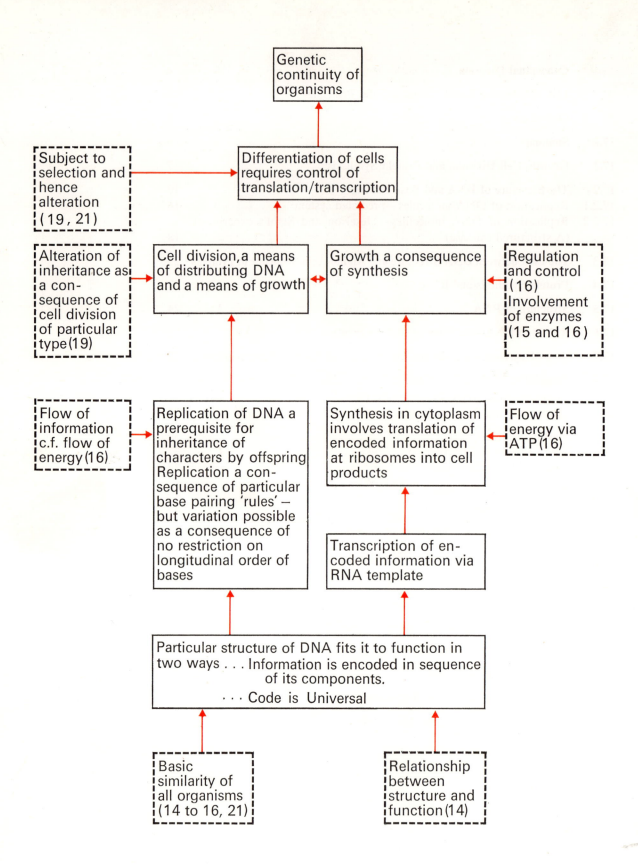

Unit numbers in brackets

Table A

List of Scientific Terms, Concepts, and Principles used in Unit 17

Taken as pre-requisites			Introduced in this Unit			
1 Assumed from general knowledge	**2** Introduced in a previous Unit	Unit No.	**3** Defined and developed in this Unit	Page No.	**4** Developed in a later Unit	Unit No.
embryo	pH	9	bacteriophage	7		
vitamin	peptide bond	13	double helix	10–18		
	primary		template	16		
	structure	13	replication	18		
	protein	13	co-linear	23		
	secondary		transcription	25		
	structure	13	mRNA	26		
	tertiary		translation	27		
	structure	13	codon	30		
	esterify	14	amino-acyl-tRNA	31		
	ultra-centrifuge	14	anti-codon	31		
	enzyme	15	rRNA	31		
	virus	15	tRNA	31		
	coenzyme	15	chromosomes	38		
			gametes	42		
			mitosis	42		
			zygote	42		
			genotype	43		
			phenotype	43		
			differentiation	45		
			switch-on/switch-off hypothesis	45		
			meiosis	77		

Objectives

1 To show knowledge of:

    (a) molecular structure of DNA   (*SAQs* 1–11, 12–23)

    (b) molecular structure of RNA   (*SAQ* 28)

    (c) differences in structure between RNA and DNA   (*SAQs* 24, 28)
        —by completing sets of multiple choice questions

    (d) replication—by giving correct bases to pair with those in a given section of DNA and RNA   (*SAQs* 1–11)

    (e) transcription—by giving correct sequence of bases in RNA given a corresponding set of DNA bases and vice versa   (*SAQ* 25)

    (f) translation—by giving correct amino acid sequence corresponding to a sequence of either DNA or RNA bases and vice versa

                                             (*SAQs* 25, 32)

    (g) main steps in synthesis of proteins—by solving simple problems in this area   (*SAQs* 25, 26, 30, 31, 35, 36, 38)

    (h) mitosis as a means of ensuring accurate distribution of genetic information   (Text questions 13–25)

    (i) genetic continuity as a consequence of the manner of replication of DNA   (*SAQs* 1–11, 33. Text questions 13–25)

    (j) variation in genetic information being related to structure of DNA   (*SAQs* 1–11)

    (k) constancy of genetic information as a consequence of specific base pairing in DNA molecules and their semi-conservative manner of replication.   (*SAQs* 1–11, 12–24, 34)

2 To analyse, evaluate, and draw conclusions from experimental data given relating to:

    (i) the manner of replication of DNA

    (ii) the role of DNA in protein synthesis

    (iii) the sequence of events in mitotic cell division.

                          (Text questions 1–8, 9–12, 13–25)

3 To apply knowledge gained during this Unit to new situations specified by test items.   (*SAQs* 27, 29, 35, 37, 39, 40)

17.0 Summary

From a consideration of the events in growth and division, two functions of DNA are revealed: self-replication and transmission of information. A more detailed treatment of the structure and chemistry of DNA follows. The relationship between structure, replication, and information-transfer leads into the subject of protein synthesis, involving DNA, RNA, and ribosomes.

The genetic code is explained, with an appendix on the methods used in this type of work.

Chromosomal behaviour during mitosis is examined, and this behaviour is linked with the distribution of genetic information from one cell generation to the next. Differentiation is briefly discussed and the switch on/off hypothesis is put forward as a partial explanation for the mechanisms of differentiation.

Some of the general conclusions and implications of molecular biology are presented.

17.1   Growth, Cell Division and Protein Synthesis

Study Comment

> The phenomena of growth and reproduction are discussed. They are complex but can be analysed in a model system provided by a virus. The structure and life history of a widely used virus is described, but you need not learn the details. Experimental work is cited and discussed which leads to the conclusion that DNA is important in information transfer from generation to generation. You should understand how this conclusion is reached from the experimental evidence.

During their lifetime all living organisms grow. This involves the production of new cytoplasm (cell material) from raw materials which the organism has captured from the world around it. Production of new cytoplasm involves the synthesis of fats, carbohydrates, and proteins, plus a wide range of other organic chemicals. Growth also involves division of cells, leading to an increase in the number of cells making up the organism. Division generally results in two daughter cells which are identical with each other and identical, except in size, with the original cell. In the case of *Protista*—the unicellular organisms—cell division leads to an increase in numbers of the organisms, i.e., in this case, cell division is a reproductive process. Growth also takes place, increasing the size of the two daughter cells after their production by division.

Production of new cell material and division of cells happen in a co-ordinated manner. Failure of co-ordination could mean, for example, the division of a parent cell before the synthesis of some important structure, and this might well seriously affect the daughter cells. So, growth and division are highly co-ordinated, even in very simple organisms (see Unit 16 for information about the control of processes inside cells). The whole process is too complicated to be describable in ordinary cells. However, in very simple organisms, the viruses, the events are few enough to be analysed in detail and described at a molecular level.

A virus is so simple that it cannot grow and reproduce on its own—it can only do so inside a living cell. Generally, each kind of virus can attack and penetrate only one specific kind of host cell. For instance the virus known as T_2 attacks the bacterium *Escherichia coli* (shortened to *E. coli*). Viruses such as this, which live inside bacterial cells, are called *bacteriophage*—or *phage* for short.

bacteriophage

T_2 phage has been widely used in investigations, partly because it is so simple, easy to keep and cheap to use, but chiefly because its reactions model the reactions taking place in more complicated cells and organisms that cannot be examined directly.

The anatomy of T_2 is known and is shown diagrammatically in Figure 1.

Chemically T_2 is made up of about thirty different proteins, plus DNA.

After infection with one or more virus particles, the host cell synthesizes new T_2 proteins and DNA—enough, within 20 minutes, to make up about 100 new T_2 'particles'.

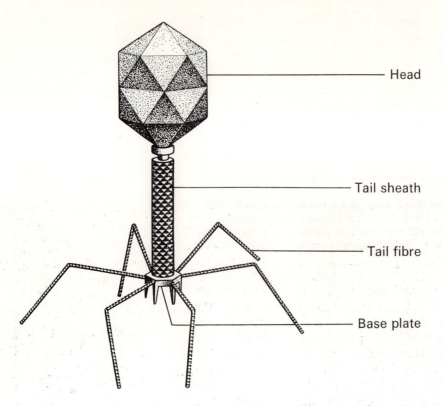

Figure 1

The anatomy of T_2 virus.

At this point the infected cell bursts, releasing the newly synthesized virus to start the cycle of infection and growth again.

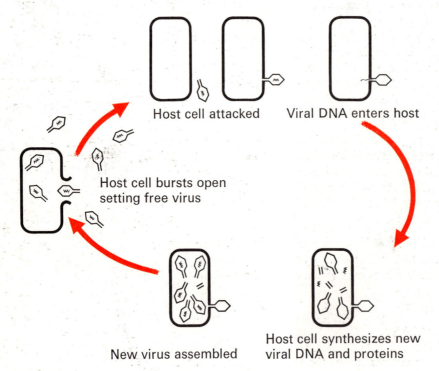

Figure 2

Life cycle of bacteriophage of T_2 type.

To perform this synthesis, *E. coli* uses its own enzymes, or makes new ones, to assemble its own amino acids into the proper sequence to make T_2 proteins and T_2 DNA. Obviously this is not the way *E. coli* normally behaves—it normally makes *E. coli* proteins and DNA. Something that

Figure 3   Two models of a DNA molecule.

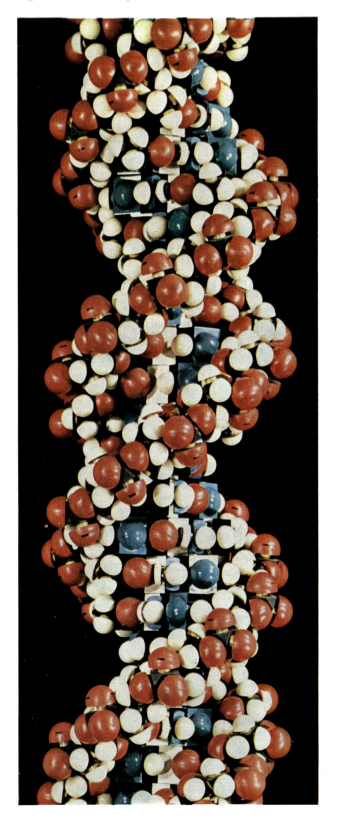 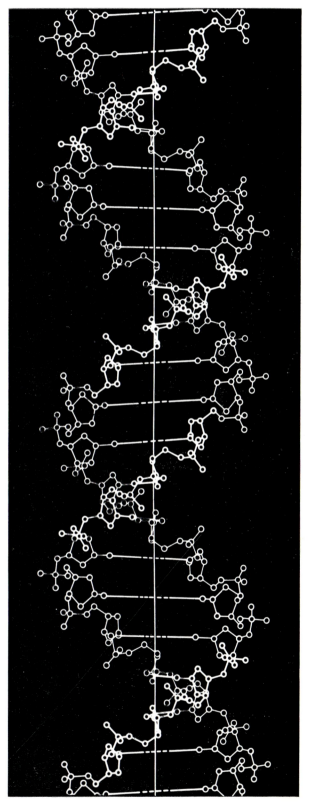

causes *E. coli* to alter its behaviour must enter at the time of infection by the virus. It could be either T_2 protein or T_2 DNA, or both. Experiments by Hershey and Chase in 1952 showed that viral protein was less likely than viral DNA to be the substance responsible for altering the behaviour of *E. coli* cells. In outline, their experiments were as follows.

They grew some T_2 virus in which the sulphur-containing amino acids of the proteins contained radioactive S atoms. and some in which the phosphate groups of the DNA contained radioactive P atoms. After the T_2 had been allowed to attack, the infected *E. coli* cells were removed from their solution, washed, and then treated in a high-speed mixer so that all the T_2 particles attached to them were broken off. Analysis showed that whereas at least 80 per cent of the ^{35}S could be recovered from the broken off material, the ^{32}P could not be. All the viral DNA had entered the host cells, and hence was not present in the solution. In spite of the violent treatment given to the infected cells, they remained capable of producing new T_2. Analysis of these new virus particles showed that they contained up to 30 per cent of the radioactive ^{32}P but very little ^{35}S.

The Hershey and Chase work supports the view that DNA rather than protein is the agent which causes *E. coli* to alter its behaviour. More recent work has settled the role of DNA. It is possible to remove the cell wall from living bacterial cells and to mix DNA extracted from phage with the naked bacteria. Under these circumstances the bacteria respond by producing fully formed and infectious virus. As no viral protein is present in the mixture, it cannot play a part in altering the behaviour of the infected bacteria. But note that fully formed viruses are produced—that is viruses with DNA enclosed in a protein sheath. The virus DNA then is the agent which determines the reproduction of the virus in *E. coli*.

To summarize, the important points are:

(a) T_2 particles are not themselves able to make more T_2 particles; this has to be done by the 'host' cell;

(b) not the whole of a T_2 particle is necessary to cause the host cell to produce T_2 particles; only the DNA is required.

The conclusion can be drawn that the T_2 DNA provides information which the host cell can use. The information must do two jobs:

roles of DNA

(1) lead to the production of new virus protein by the host cell;

(2) lead to the production of new T_2 DNA; that is, cause the information-carrying DNA itself to be reproduced or replicated.

DNA can be found concentrated in the nucleus of cells of more complex living organisms. This is true of every type of cell, including those which take part in sexual reproduction—egg and sperm cells. The link between DNA and the transfer of genetic information is now firmly established. As a result, a good deal of attention has been paid to the chemical structure and behaviour of the DNA molecule.

17.2   The Structure of DNA and Base Pairing

Study Comment

The chemical and physical structure of DNA is described. Do not attempt to learn the details of either its structure or chemistry. You should, however, remember that DNA is a giant molecule and contains four kinds of subunit which are linked in the molecule in a particular way. Because of its particular structure, DNA could fulfil the two functions described for it in section 17.1. This point is discussed in 17.2.1 as far as replication is concerned. In 17.2.2 experimental evidence is cited which supports the method of replication described in 17.2.1. The experimental work and the conclusions are presented as a structured exercise, to give you practice in handling data and drawing deductions. You should remember how DNA replicates, and enough details to understand the relationship between structure and function.

You will be familiar with some aspects of the chemistry and structure of DNA already from Units 13 and 14. In this section, aspects of DNA structure related to its role in cells are examined in more detail. You will find an alternative account to the one given here in *The Chemistry of Life*, pp. 70–76.

Look at the photograph and diagram of models of DNA—it is a complex molecule. See Figure 3, facing p. 8.

  (1) It has a regular shape—the same width all the way up—and could plainly be of any length, like a piece of rope, a chain, or a spring.

  (2) Like a piece of rope, it has a regular twist.

  (3) If we start to take it apart we find that, also like a piece of rope, it consists of two strands which coil around each other. The whole molecule can be described as a double helix.

Do not attempt to remember the full chemical composition or formula of DNA—note only the diagrammatic formulae.

  (4) We can also take it apart from one end instead of from side to side, and then we find that, like a chain or a pile of bricks, it is a stack of endlessly repeating units, piled on top of one another.

The structure of these units explains how the regular-looking molecule is built, and why it can be taken apart in this way.

There are two kinds of *unit*.

One looks like this:

this can be split into:

and

It is a pair of unequal subunits called nucleotides.

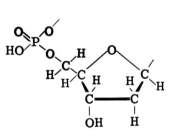

Each subunit has

(1) a deoxyribose phosphate ester shown as: or

(2) a base shown as: or   The smaller rectangle represents a pyrimidine, and the larger a purine.

deoxyribose phosphate ester

base

and

base

deoxyribose phosphate ester

The two bases represented here are thymine (T), a pyrimidine, and adenine (A), a purine. (S = sugar)

The other unit has exactly the same shape, but its bases are different: cytosine (C) (a pyrimidine) and guanine (G) (a purine).

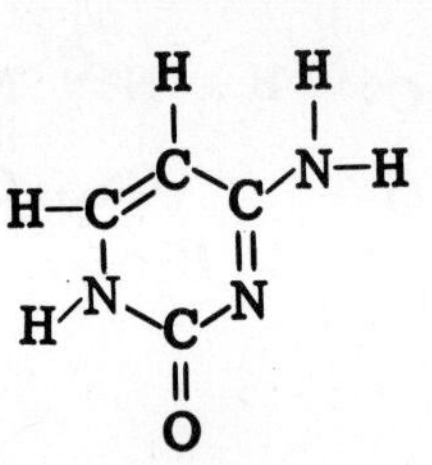

cytosine

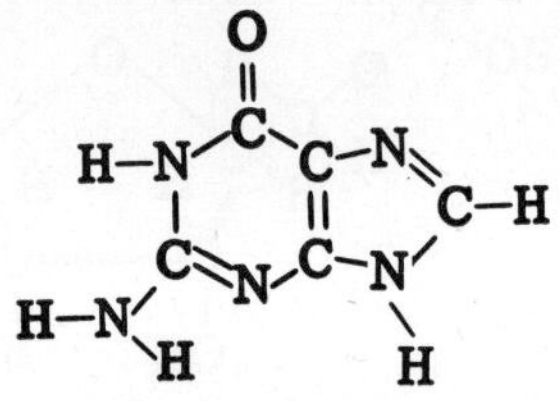

guanine

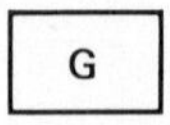

So by changing the bases in the subunit, four nucleotides can be made.

The sugar phosphate groups are the same, however,

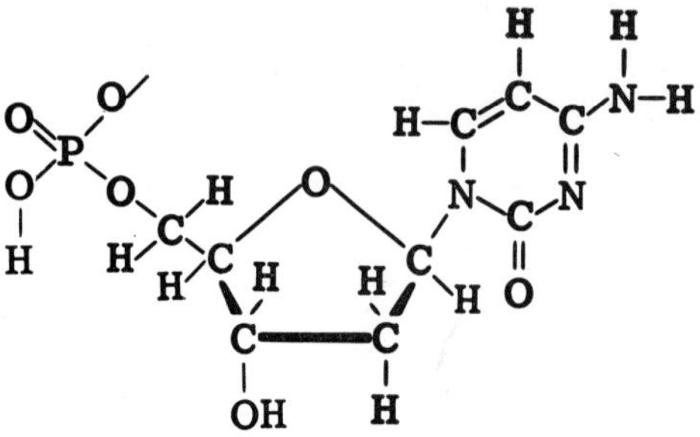

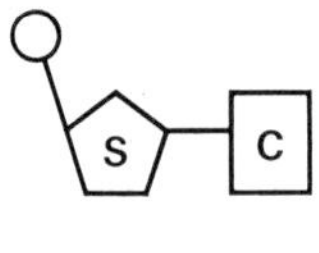

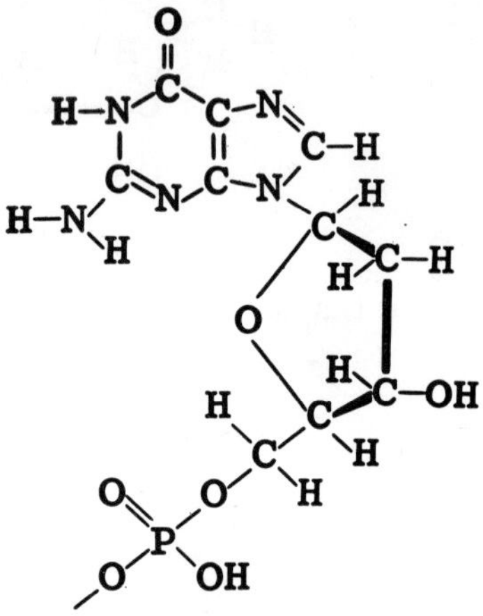

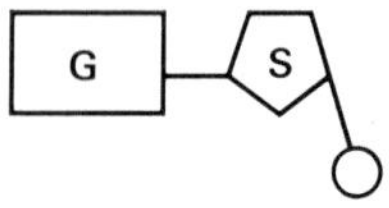

and these subunits bond to each other to produce a unit the same shape as the A–T unit.

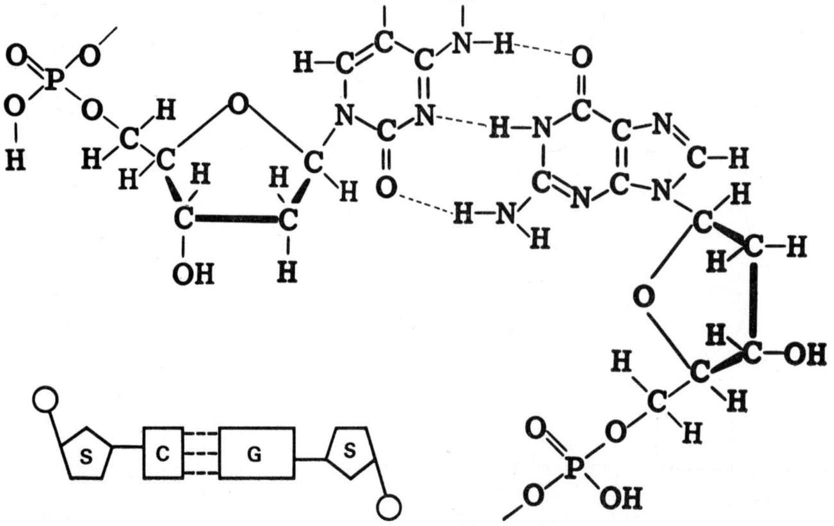

C–G units and A–T units can be stacked together and their overall shape remains unaltered,

even when turned round.

So they can be stacked in any order.

In the intact DNA molecule, each unit is joined to the one above and the one below through the phosphate groups, which esterify with the sugar molecules.

Ester bonds are covalent and very strong; they provide a firm backbone for each half of the molecule. The two backbones twist around each other, spaced the right distance apart by the sugars and bases in between them, and held together by the bonds between the bases.

These bonds between bases are *hydrogen bonds*, formed by the sharing of a hydrogen atom between two electronegative atoms, oxygen and nitrogen or nitrogen and nitrogen. They are much weaker than the covalent bonds of the backbone (see Unit 12).

13

The difference between bond strengths is shown in Figure 4.

So the two strands can be pulled apart much more easily than the stacked units can be unstacked or degraded. Heating DNA samples from some organisms to 80°C, followed by slow cooling or treatment with alkali, will separate the strands. And when the temperature or pH is lowered, the two strands will come together again with the same bases paired. No other base pair arrangements can be made which will form units of exactly the same size and conformation.

This can be seen in Figure 5. The proper pair of bases are shown in the left-hand diagram—adenine/thymine; but on the right, adenine is paired with either adenine or guanine. Should this actually happen, the distance between the backbones of the molecule would have to be increased to fit the A/A or A/G subunits, which are larger than A/T or G/C subunits.

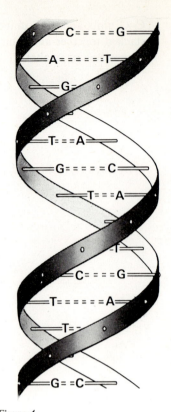

Figure 4

DNA double helix showing the two types of bonding – stronger bonds in the backbone region and weaker bonds linking the two strands together.

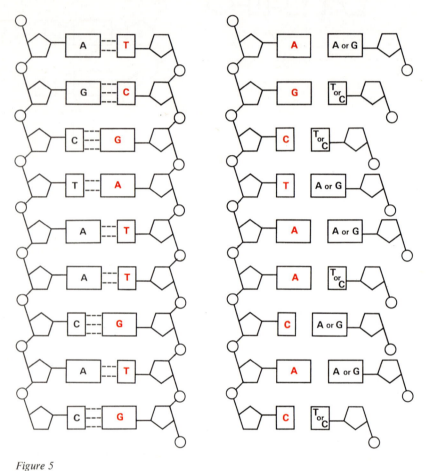

Figure 5

'Legitimate' base pairing.   *Arbitrary base pairing in DNA.*

Similarly a base pair of T/T, C/C, or T/C would only fit into the molecule if the distance between the backbones was made smaller at this point. There is strong evidence that the distance between the backbones does not vary along the length of the molecule. When this distance is measured it turns out to be the distance which fits a purine-pyrimidine pair.

There is a further limitation on base pairing, apart from the alteration in size of the unit which would follow from illegitimate pairing. Whether, for example, adenine can form a hydrogen bond with cytosine or thymine depends upon the position of hydrogen atoms in the prospective pair.

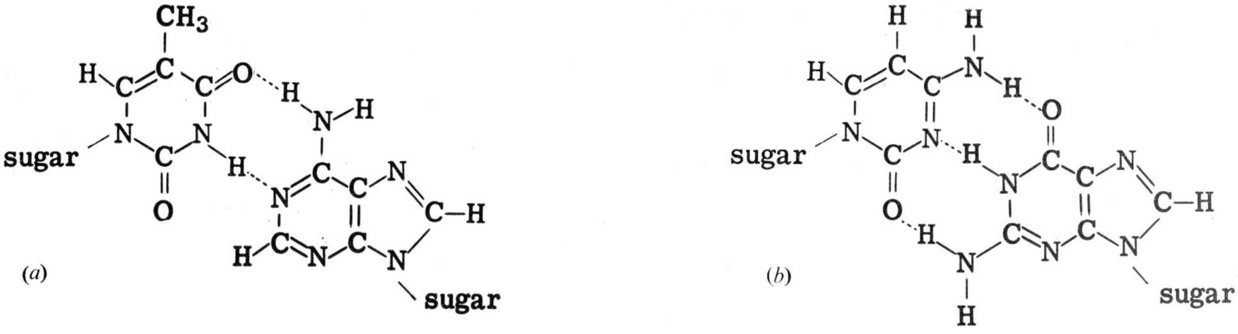

Figure 6

Base pairing by hydrogen bonding between (a) adenine and thymine and (b) cytosine and guanine.

The arrangement in which thymine bonds with adenine, and cytosine bonds with guanine, not only fits the space between the parallel backbone of sugar phosphates but also allows the greatest number of hydrogen bonds to form between the pairs of bases (Fig. 6). Although these are weak chemical bonds, they are very important in determining the stability of the whole assembly. If thymine and guanine are paired then only two hydrogen bonds can be formed between them, as can be seen in Figure 7.

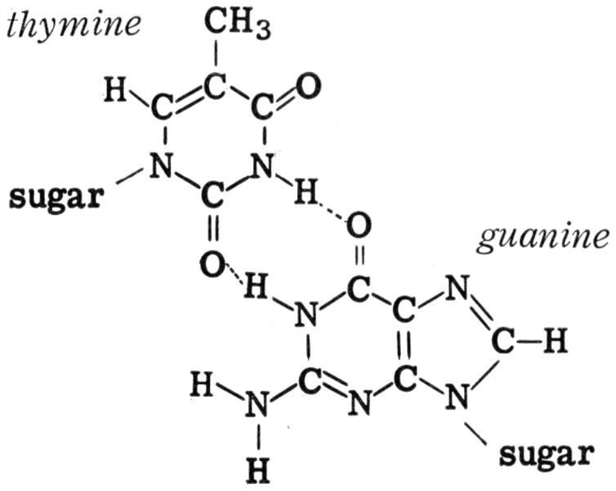

Figure 7

To show three alternative arrangements in which thymine and guanine might bond together.

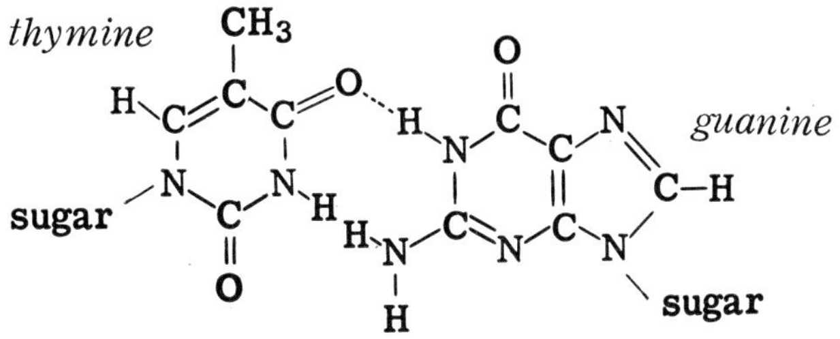

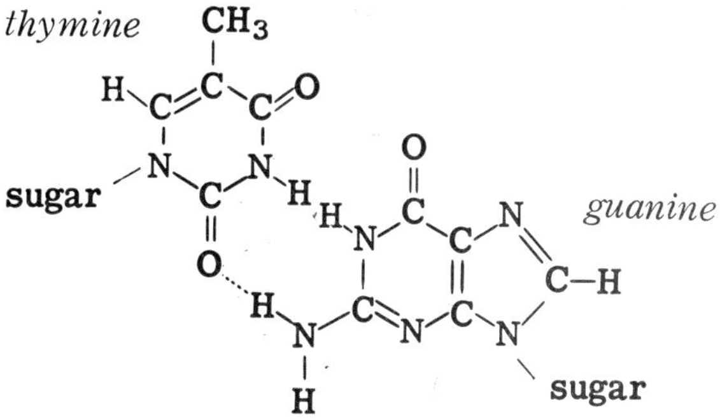

Similarly, only one hydrogen bond can form between adenine and cytosine if these two bases are paired. Consequently a DNA molecule in which thymine paired with guanine, and cytosine with adenine would be less stable than one in which thymine paired with adenine, and cytosine paired with guanine.

In addition, in none of the possible illegitimate pairings would the bonds to the sugar molecules be in the same relative position as they are in the standard or legitimate pairings. In other words the shape of such a unit would not enable it to fit into the double helix without distortion.

Chemical analysis of DNA from a wide variety of organisms also suggests that the pairing is adenine with thymine, and guanine with cytosine. For it is found that the relative amount of adenine equals that of thymine, and that of guanine equals that of cytosine, in any one sample. But there is not this equality between cytosine and adenine, nor between guanine and thymine. This can be seen from the data in Table 1.

Table 1 Relative amounts of the four bases in DNA from various sources

Organism	Adenine	Guanine	Thymine	Cytosine
Escherichia	25.4	24.1	24.8	25.7
Yeast	31.3	18.7	32.9	17.1
Salmon	29.7	20.8	29.1	20.4
Mouse	29.7	21.9	25.6	22.8
Wheat	26.8	23.2	28.0	22.0
Man	30.4	19.6	30.1	19.9

17.2.1 Replication of DNA molecules—predicted system

As a consequence of strict base pairing, if the two strands of a DNA molecule are separated from one another, each strand can serve as a set of instructions—a 'mould' or a 'template' for making another strand to fit it; i.e., it replicates.

The new strand formed on this template would be exactly like the strand just removed from it, so the new DNA molecule would be exactly like the original.

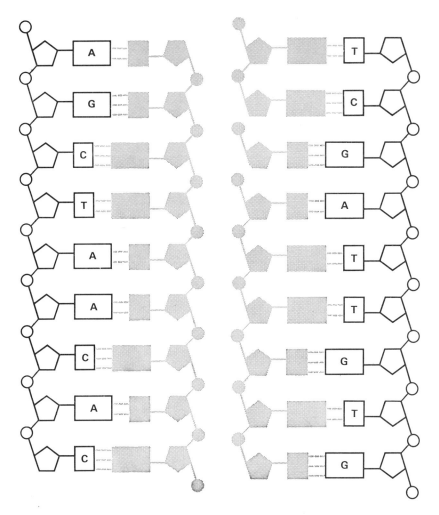

Notice that in the original molecule, if you read the bases in the left-hand strand from the top, they go A–G–C–T–A–A–C–A–C. In the two molecules just made, the same strands read the same. In fact, one of each pair of strands *is* the same strand as in the original molecule, but the other consists of an entirely new set of bases and sugar-phosphate molecules.

Thus, from what we have seen of its chemical structure, the DNA molecule *could* perform the functions that experiments (see 17.1) have suggested for it. These are:

(1) the carrying of coded information regarding protein synthesis;

(2) the reproduction and transfer of this information from one generation to the next.

In the first case, since the base pair units of the chain can be stacked in any order, they could make up a kind of 'alphabet' which could be used to construct 'words' and 'sentences'. By this method an enormous amount of information could be encoded in the molecule.

In the second case, once a molecule with a particular order exists, that order can be preserved and reproduced indefinitely and exactly by separating the two backbones and making use of the exposed bases on each of them to construct new complementary chains. The exactness of the copy depends on the regularity and consistency with which adenine and thymine pair only with each other, and guanine and cytosine only with each other.

The structure of the DNA molecule fits the data from a variety of investigations into the composition, size and shape of these molecules. Once the double helical structure had been put forward by Watson and Crick in 1953, it was realized that this same structure provided a possible way for DNA replication to occur. But just as hard facts support the three-dimensional structure proposed for the molecule, so hard facts are needed to support the manner of replication put forward. It is not enough to see that DNA *could* replicate in the manner suggested by its structure—it is necessary to show that in fact it *does* replicate in this way.

(Note: you might care to read Watson's account of how he and Crick came to propose the double helix structure in *The Double Helix*, by J. D. Watson. This is not essential reading, but it is fascinating, exciting, and very different from conventional accounts of scientific discovery.)

17.2.2 Replication of DNA molecules—Meselson and Stahl's experiment (A structured exercise)

First consider the following points about the DNA molecules.

(1) It is known that DNA is a linear molecule made up of two strands.

(2) The molecule is composed of molecular subunits that contain nitrogen atoms.

(3) During cell division, the mother cell divides in such a way that the two daughters each contain as much DNA as did the mother cell.

(4) The process of replication must be such that the daughter cells receive genetic instructions similar to those of the mother cell.

In general there appear to be three ways in which the replication could be performed. These are shown diagrammatically.

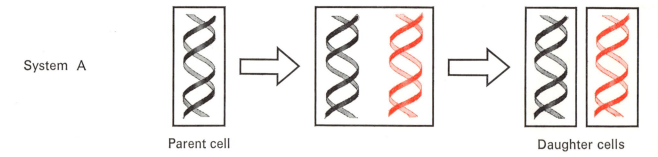

System A

Parent cell

Daughter cells

Conservative System

The two new DNA strands occur together in one daughter cell, the two old in the other.

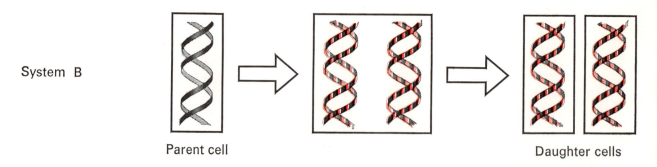

System B

Parent cell

Daughter cells

Dispersive System

New and old DNA occur together in both daughter cells. The old DNA is dispersed amongst all the strands in the daughter cells.

System C

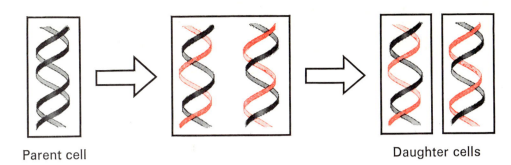

Parent cell   Daughter cells

Semi-conservative System

New and old DNA occur together in both daughter cells but old DNA and new DNA are each in their own strands.

Two American biologists, Meselson and Stahl, devised a critical test which enabled them to select one of the three replication systems as being the correct one. They made use of three techniques.

 (1) They grew cells in a medium which contained ^{15}N, enabling them to distinguish newly formed DNA from parent DNA.

 (2) They were able to separate the two strands of DNA by a mild chemical treatment.

 (3) By using a density gradient in an ultracentrifuge, they were able to measure the density of DNA strands. The technique is outlined in Unit 14.

Meselson and Stahl grew bacterial cells in a ^{15}N enriched medium for many generations. Then a sample of the culture was put into a new medium which was not ^{15}N enriched. As normal nitrogen, ^{14}N, is lighter than ^{15}N, the DNA made by the cells in the ^{14}N medium would have a lower density (i.e., weight per unit length) than the DNA made while the cells were in the ^{15}N. Strands of DNA of different density can be separated from each other in a density gradient centrifuge (described in Unit 14).

Question 1*

If the two strands of DNA in each daughter cell were separated and all were suspended in a density gradient, would all the strands have the same weight in System A?
        System B?
        System C?

Question 2

In what proportion would you expect heavy and light strands after one replication in each of the three systems?

---

* Check your answers with those provided on p. 71.

The experimental results obtained by Meselson and Stahl are shown diagrammatically as follows:

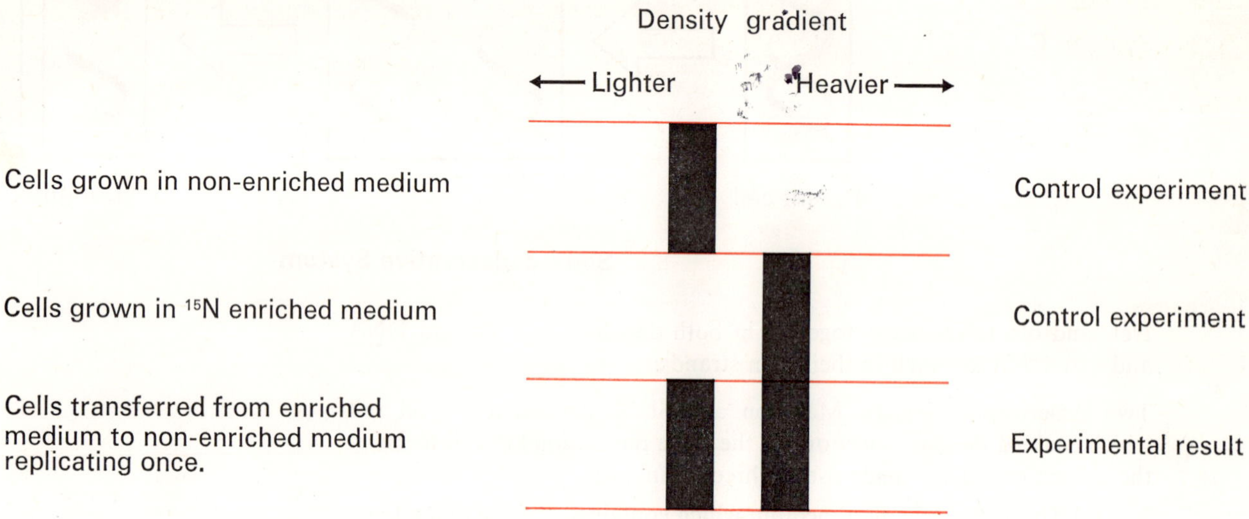

Density gradient

←— Lighter   Heavier —→

Cells grown in non-enriched medium   Control experiment

Cells grown in ^{15}N enriched medium   Control experiment

Cells transferred from enriched
medium to non-enriched medium
replicating once.   Experimental result

Note

    (i) In the diagrams, the *width* of the bands is a measure of the amount of DNA present, and their *position* shows the relative density of DNA in the sample examined.

    (ii) The DNA is denatured—the double strands are separated before their density is measured.

Question 3

Can you decide which of the three replication systems proposed accounts best for the experimental results?

Question 4

Do the experimental results show that any of the systems are correct?

Question 5

What firm conclusion, if any, can you draw from the data provided from Meselson and Stahl's experiments?

When whole DNA molecules were centrifuged in a density gradient, a rather different result was obtained. The DNA in these experiments was not separated into its constituent strands before centrifugation.

Question 6

On the diagrams (p. 21), fill in the results you would expect to obtain after centrifuging whole (i.e., double-stranded) DNA from cells grown first in ^{15}N and then for one generation in ^{14}N.

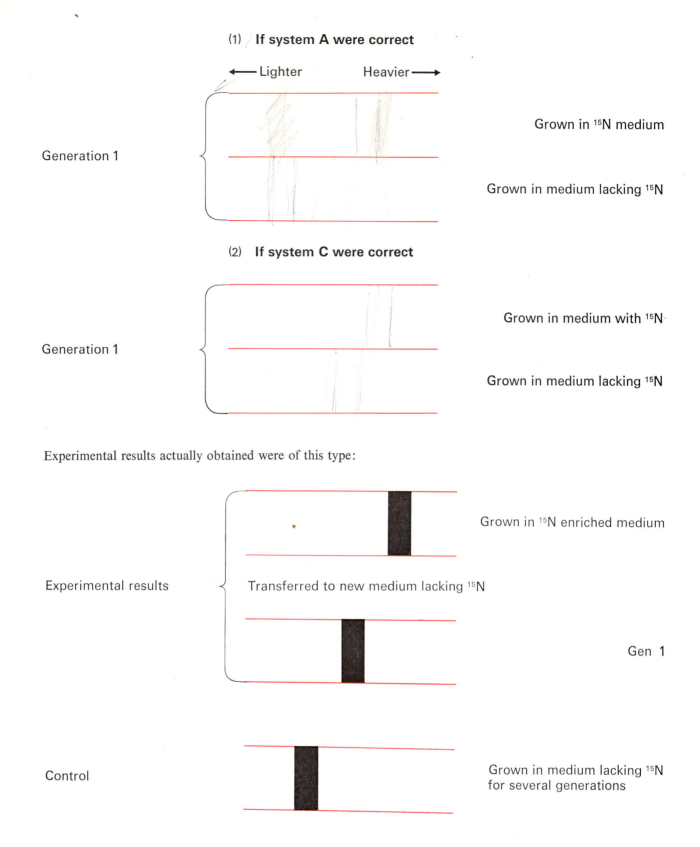

(1) **If system A were correct**

←—— Lighter   Heavier ——→

Generation 1

Grown in ¹⁵N medium

Grown in medium lacking ¹⁵N

(2) **If system C were correct**

Generation 1

Grown in medium with ¹⁵N

Grown in medium lacking ¹⁵N

Experimental results actually obtained were of this type:

Experimental results

Grown in ¹⁵N enriched medium

Transferred to new medium lacking ¹⁵N

Gen 1

Control

Grown in medium lacking ¹⁵N
for several generations

21

Question 7

Which of the replication systems best fits these experimental results?

By extending the experiments, to examine cells grown first in ^{15}N and then 1, 2, and 3 generations in ^{14}N medium, the following results were obtained.

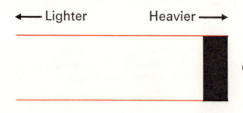

Grown in ^{15}N enriched medium

Cells transferred to medium lacking ^{15}N

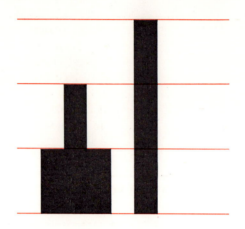

After 1 generation

After 2 generations

After 3 generations

Control

Grown in medium lacking ^{15}N

Question 8

How may these results be explained?

Check your answers against those provided on pp. 71–73.

Now answer *SAQs* 1–11.

17.3 The Link between DNA Structure and Amino-acid Sequence in Proteins

Study Comment (sections 17.3, 17.4, and 17.5)

> The manner in which DNA and protein synthesis are related is discussed (17.3). It is shown that besides DNA, another kind of nucleic acid, RNA is involved in transfer of information from the nucleus to the cytoplasm, where protein synthesis occurs (17.4). The structure of RNA is described in 17.5. Do not learn the detailed chemical composition of RNA.

As the experiments described in section 17.1 show, the transfer of information from parental generation to the progeny involves the transfer of information leading to the production of new proteins rather than the transfer of proteins which can be copied (see Unit 16 for discussion about information flow in cells). Proteins are linear polymers of amino acids assembled in particular sequences. The particular sequence has a great bearing on the chemical and physical properties of the protein, as you saw in Unit 14. The fact that information transfer is by DNA—a long molecule which itself is made up of a sequence of subunits—suggests that the information governing the sequence of amino acids in a protein is contained in the sequence of bases in DNA. The simplest model of this is one in which DNA and protein are co-linear; that is the linear sequence of amino acids in a particular protein corresponds directly to the linear sequence of bases in a particular stretch of DNA. If this hypothesis were true it is clear that if the sequence of bases in DNA is altered there ought to be an alteration in the sequence of amino acids in the protein produced.

One test of this hypothesis was made by comparing the amino-acid sequence normally produced from a DNA molecule with that produced from a similar DNA molecule in which the sequence of bases had been altered, deliberately or naturally.

Because the techniques used in investigations of this kind are complex and the design of the experiments sophisticated, it is not possible to describe them here. The results obtained have, however, convinced biologists working in this field that the DNA and proteins are co-linear. Two major questions now need answers. First, just how is the assembly of protein carried out and, second, just how is the information carried? What, for example, is the unit of DNA which corresponds to one amino acid in a protein?

17.4 Protein Synthesis and RNA

Assembly of amino acids into proteins occurs in the cytoplasm of living cells, whereas the DNA of the cell occurs very largely in the nucleus. Clearly, then, the assembly of proteins cannot take place as the result of particular amino acids reacting directly with particular groups of bases along the nuclear DNA. Any idea that DNA itself is a *template* for protein assembly must be incorrect. Experiments show that protein synthesis occurs in the cytoplasm at the ribosomes. (See *Microstructure* and Unit 14.) During protein synthesis, ribosomes are associated with a long strand-like molecule that is composed of a substance called RNA (ribonucleic acid).

17.5   Structure of RNA

This is very similar to DNA.

(1) It is a chain-like molecule of stacked units.

(2) However, RNA consists of only one strand, not two coiled round each other, as they were in DNA.

(3) As in DNA, the sugar base units are covalently linked through phosphate groups to provide a strong backbone, but the sugar is different. It is ribose, not deoxyribose.

ribose   *deoxyribose*

(4) Three of the bases are exactly the same—adenine, guanine, and cytosine. However, in RNA, thymine is entirely replaced by uracil, but note that uracil is a pyrimidine derivative and can form hydrogen bonds with adenine, in the same way that thymine does in DNA.

thymine   *uracil*

17.6   RNA and DNA – the Process of Transcription*

Study Comment

> In the process of transferring information from the nucleus to the cytoplasm, information carried in the DNA molecules is transcribed into RNA molecules. This process is outlined.

In spite of these differences, RNA is so similar to DNA that a single strand of DNA is able to pair with and coil around a strand of RNA, as long as the base pairing rules are satisfied. In other words, the two strands are complementary, like the two strands of a DNA duplex coil. In such an DNA and RNA 'hybrid' duplex, adenine and uracil must always be paired, and guanine must always be paired with cytosine.

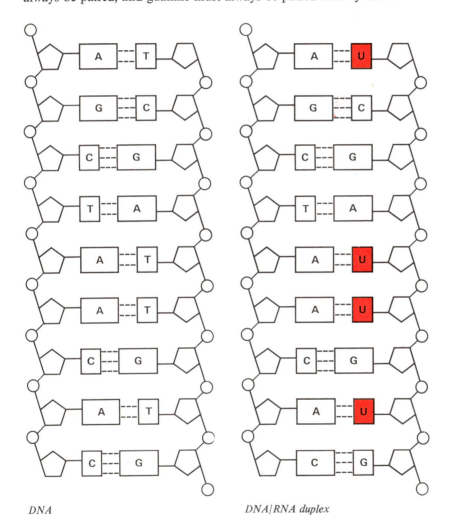

DNA   *DNA/RNA duplex*

Because the base pairs are formed in the same way, the 'message' (that is the base sequence) of one DNA strand can be copied or transferred into an RNA strand. In cells this is done with the help of an enzyme, RNA polymerase, which synthesizes RNA in the presence of DNA. The RNA strand then leaves the DNA strand and is used as the message or information for making up the particular sequence of amino acids that will form a protein.

* Describes the process by which information encoded in DNA is converted into information encoded in mRNA. More generally transcription refers to the same language with different accents, compared with translation, which refers to two languages (see p. 27).

The single-stranded RNA copies of DNA are called *messenger RNA,*
mRNA for short

    (1) DNA makes DNA (replication).

Now answer *SAQs* 12–24, pp. 54–57.

(2) DNA makes RNA (transcription).

17.7 Some Experimental Evidence for the Role of RNA in Transcription

Study Comment

> **Experimental data is given which supports the view that RNA is involved in information transfer. The details of the experiments need not be remembered, however.**

The synthesis of a particular protein molecule requires the assembly of particular amino acids in a unique sequence. The information which specifies both the amino-acid molecules and the order in which they are to be assembled is carried in a coded form by the sequence of bases in the part of the DNA molecule concerned with the particular protein. In the system just described, the transcription of DNA-coded information into RNA-coded information results in the synthesis of a strand of mRNA which matches the DNA strand. It is this mRNA which is used for translation* of the coded information into the amino-acid sequence of the protein molecule.

translation

If we are to believe this description, it must be demonstrated experimentally that mRNA can in fact determine an amino-acid sequence. For clearly, if mRNA stands between DNA and the translation process, the ability of mRNA to determine amino-acid sequences is vital to the model. It has been possible to test this particular ability of mRNA by using a virus as the test organism. The particular virus used is one which causes disease in tobacco plants. Two strains of this virus each cause quite distinctive and characteristic symptoms in their host plant. These particular viruses do not contain DNA but instead consist of a protein sheath surrounding a central core of RNA. It was found possible to separate the central RNA core from the protein sheath in each case. When separated cores and sheaths were mixed together, some reassembly took place, so that some fully functional virus was reconstituted from the separate parts. Further, if RNA core from one strain, A, was mixed with sheath from the other virus, *B*, active virus could be obtained—activity being decided by the ability of the virus to cause disease in healthy tobacco plants. Tested in the same way, RNA cores alone caused disease but sheath material alone did not. The mixture of strain-B core RNA and strain-A sheath material also produced some active virus. Using separated sheath material it was possible to measure the amount of each amino acid in the sheath protein of both strains. The proteins were found to be distinctly different from each other. When the test plants were inoculated with virus particles, they developed disease symptoms; new virus particles were then harvested from the diseased hosts and examined and analysed.

Results from a series of such experiments carried out by Fraenkel-Conrad and Singer in 1957 are shown in the Table 2, p. 28.

Before looking up the table of results, answer the following questions. After you have answered each question, check your answer against those provided on p. 73.

---

* Describes the process by which the encoded information of mRNA is decoded and 'rewritten' as a sequence of amino acids.

Question 9

If the new generation of virus made protein-coat material similar to that of their parents, then the information controlling this could have been carried in either:

(a) protein coat of parent; or

(b) RNA of parent.

If (a) is correct, then what sort of protein would you expect to find made by the progeny of virus strain-*A* RNA core/virus strain-*B* sheath?

If (b) is correct, then what protein would you expect to be made by progeny of virus strain-*B* RNA core/virus strain-*A* sheath?

Question 10

If both the protein sheath *and* RNA core play a part in determining the composition of the protein made by the progeny, what type of protein would you predict would be made by virus strain (core RNA B/sheath protein A)?

Now examine the table of results:

Table 2

Virus used to infect host plant	RNA from strain	A	A	B	B
	Protein sheath from strain	A	B	A	B
Symptoms of disease in host		A	A	B	B
Per cent weight of amino acids in protein from harvested virus progeny.	Glycine	2·3	2·3	1·8	1·6
	Alanine	6·5	6·9	8·5	8·6
	Serine	9·0	8·8	8·1	8·1
	Methionine	0	0	2·2	2·0
	Histidine	0	0	0·7	0·7
	Tyrosine	4·1	4·3	6·2	6·3
	Lysine	1·9	1·8	2·3	2·4
	Glutamic acid	12·4	12·1	17·3	16·4
	Valine	9·6	9·0	6·3	5·9

Question 11

If both sheath and RNA play a part in determining the structure of the protein made by the progeny, could the absence of the amino acids methionine and histidine from the protein made by the progeny of parent strain (type A RNA/type B sheath protein) be explained?

Question 12

Is there any evidence to support the view that the structure of sheath protein in the new generation is determined by the parental protein?

If the conclusions from this investigation are taken together with the conclusions from the experiments considered in section 17.1, it seems that the protein is determined by the particular structure of nucleic acids and not by proteins themselves, i.e., it is the nucleic acids that are the vehicles carrying inherited genetic information.

The investigations of Fraenkel-Conrat and Singer show that RNA can, at least in the situation which they used, specify the order in which amino acids are assembled. Other experiments have confirmed the belief that RNA stands between DNA and the translation process. For example, it is possible to make cell homogenates which are free of DNA and to obtain some incorporation of radioactive amino acids into proteins in the homogenate. No protein synthesis will occur, however, unless RNA is present in the system. It is even possible to replace the natural RNA in the homogenate by synthetic RNA to obtain polypeptide synthesis.

Consequently, it seems possible that RNA can play the role predicted for it (in 17.4) in the transcription and translation processes described in this section.

17.8 Translation of the RNA Message into an Amino-Acid Sequence

Study comment

Information carried in both DNA and RNA is carried in a coded form. The code is written in nucleic acid bases.

Some of the characters of the code language are described and should be learnt. In the decoding of the information in the cytoplasm, two other types of molecule are involved; a new kind of RNA (tRNA), and amino-acid activating enzymes. The roles of these two substances in translating the encoded information are outlined in Section 17.8.1 and should be learnt. Translation also involves the ribosomes of the cell, as outlined in 17.8.2.

You already know:

(1) that the sequence of bases in a length of DNA determines a complementary sequence of bases in mRNA which determines a corresponding sequence of amino acids;

(2) that there are only four bases in DNA, but there are twenty different amino acids that can occur in proteins.

So each amino acid must correspond not to one base of the mRNA chain but to a group of bases, in fact to a group of three. The argument for this runs along the following lines.

If the size of the coding group in DNA were one base, then only four amino acids could be coded. Hence the coding group must be larger than one base. With two bases in the coding group, sixteen different pairs are possible, as shown in the chart below.

		Second base of group			
		C	G	A	T
First base of group	C	1	2	3	4
	G	8	7	6	5
	A	9	10	11	12
	T	16	15	14	13

Again there would be too few codings to allow each of the twenty amino acids to correspond to one different group.

If the coding group were a group of three bases, then there would be $4^3 = 64$ different possibilities; and if a group of four, then there would be $4^4 = 256$ possibilities. So the group must be a minimum of three bases, but of course could be larger than this, or of variable size. Experimental results support the hypothesis that the group size is three bases per amino acid. This group of bases is called a *codon*.   codon

With 64 different codons available to code for 20 amino acids, there could be more than one codon for each amino acid and/or some codons which are 'nonsense' codons and do not code for an amino acid at all. This is discussed in more detail in 17.11.

It is possible that an amino acid could be directly matched with a codon of the mRNA. However, this does not appear to be so. Instead an adaptor molecule plays a part. This has the property of attaching to one particular amino acid, and then 'recognizing' the codon in the messenger RNA which corresponds to that particular amino acid.

17.8.1 Translating the RNA message—the role of tRNA

The adaptor molecules are also made of RNA. There is at least one kind of adaptor molecule for each amino acid. Each adaptor can become attached to only one kind of amino acid. We do not know exactly what any adaptor molecule looks like, but we do know the sequence of bases in a number of them. We also know that the adaptor RNA molecules are double-stranded like DNA, for most of their length. Their sequence of bases suggests that they fold up into a secondary structure rather like a clover-leaf.

The adaptor molecules are called *transfer RNA* or *tRNA for short*.

      tRNA

Attachment of the amino acid is done by an enzyme called an *amino-acid activating enzyme*. There is a different activating enzyme for each different amino acid.

Once joined, the whole complex, called an *amino-acyl-tRNA*, is ready to recognize a codon on the messenger RNA. This recognition is possible because the 'loops' of the clover-leaf contain short sequences of bases that are unpaired. A short segment of one of these sequences in the central loop is complementary to a codon in the messenger RNA, and can pair with it, exactly as bases pair on DNA duplexes. This recognition segment is called the *anticodon*.

      amino-acyl-tRNA

      anticodon

The process of translation from mRNA into proteins is specific if both:

    (1) the binding of the amino acid to the tRNA;

    (2) the recognition of the amino-acyl-tRNA complex by messenger RNA, are highly specific. The first occurs because each amino acid has its own activating enzyme and the second is due to recognition between each mRNA codon and its own anticodon. Recognition occurs while both mRNA and the tRNA adaptor are attached to a structure called a ribosome.

17.8.2 Translating the RNA message—the role of ribosomes

The ribosome is a complex structure consisting of two subunits of slightly different size. Each subunit contains 30–50 protein molecules and one molecule of RNA. This RNA is called *ribosomal RNA* or *rRNA*.

      rRNA

There are thousands of ribosomes in nearly every living cell—a reflection of the importance of protein synthesis—for they are essential to the process. The way in which the mRNA and the amino-acyl-tRNAs interact with ribosomes is shown in the diagram in the next section.

17.9 Protein Synthesis

Study comment

This section is concerned with describing the whole process of protein synthesis on ribosomes. The roles of each component in the system are illustrated. Do not attempt to learn the details, only the main features of the process. An account of protein synthesis is given in *The Chemistry of Life*, pp. 159–174, which you could read if you have time.

17.9.1 Transcription and translation

The stages can be described as:

(1) mRNA binds to the smaller ribosome subunit.

(2) An amino-acyl-tRNA which has a group of bases complementary to a group in the messenger binds to both parts of the ribosome in position P. But, in order to start synthesis of a whole polypeptide chain, a tRNA carrying a special amino acid, *n*-formyl methionine and having an anticodon complementary to AUG must bind to the ribosome at position P. The next tRNA molecule approaches the other binding site at position A.

(3) The second amino-acyl tRNA binds to the ribosome in position A, recognizing the next codon in the messenger. In the diagram, the next codon is G–C–C, and the tRNA which recognizes that, always carries the amino acid alanine.

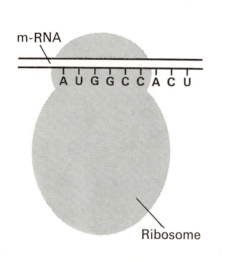

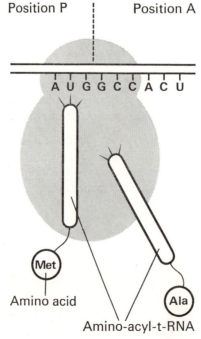

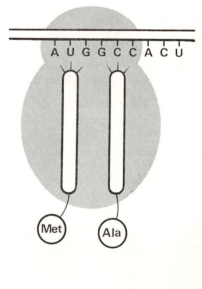

(4) The two amino acids are now next to each other and a reaction takes place between them to form a 'peptide bond', the bond which connects amino acids to form a polypeptide or a protein chain.

(5) A reaction releases the tRNA from site P. The second tRNA, which now has both amino acids attached to it, moves from site A to site P. A third amino-acyl-tRNA, threonyl-tRNA, carrying the amino acid threonine, recognizes the next messenger group of bases, A–C–U, and attaches to site A which is exposed by movement of the mRNA relative to the ribosome.

(6) The sequence is repeated as from stage 3.

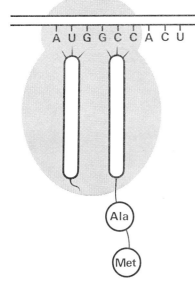

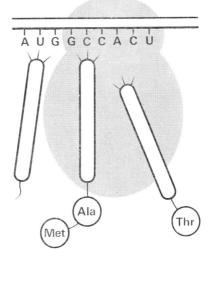

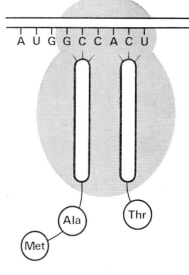

(7) This series of events ends when one of the 'stop' codons occurs in the mRNA (see 17.11 for details). No further amino acids are added to the polypeptide chain. Instead, both it and the tRNA separate from the ribosomes and each other. The polypeptide is now free in the cytoplasm of the cell.

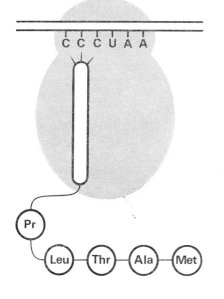

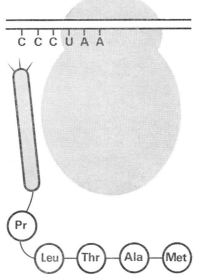

33

17.9.2 From nucleus to cytoplasm

The various stages in protein synthesis are summarized below.

(1) Transcription of specific segments of DNA into mRNA molecules in the nucleus.

(2) Movement of mRNA and tRNA from the nucleus into the cytoplasm. How this takes place is not known.

(3) Formation of amino-acyl-tRNA and binding of mRNA on to ribosomes. Formation of amino-acyl-tRNA requires ATP.

(4) Reactions as described in translation sequence.

(5) Release of the completed polypeptide chain from RNA and ribosomes. This probably requires ATP.

(6) Folding of polypeptides into their final shape.

17.10   The DNA and RNA Code

Study comment

> A more detailed description of the coding of DNA and RNA is given. Again do not attempt to learn the details, but do learn the main features of the code discussed in section 17.10.1.

You have seen that, by determining the order of the groups of three bases in the nucleic acids, the order in which amino acids become assembled into chains is also determined.

The three base groups constitute a 'code' for the amino acids, and the whole assembly of such groups makes up *the genetic code*.

The code has been deciphered completely in the case of mRNA. It turns out to be very interesting, and it can be written in the form of a table (Table 3). It is of course complementary to the triplets in one strand of DNA.

The black-page Appendix 1 gives more information on how the code was broken (p. 49).

17.10.1   The mRNA code

Table 3   **mRNA Codons**

First Letter		Second Letter				Third Letter
		U	C	A	G	
U	U	UUU ⎫ phe UUC ⎭ UUA ⎫ leu UUG ⎭	UCU ⎫ UCC ⎪ ser UCA ⎪ UCG ⎭	UAU ⎫ tyr UAC ⎭ UAA stop UAG stop	UGU ⎫ cys UGC ⎭ UGA stop UGG try	U C A G
	C	CUU ⎫ CUC ⎪ leu CUA ⎪ CUG ⎭	CCU ⎫ CCC ⎪ pro CCA ⎪ CCG ⎭	CAU ⎫ his CAC ⎭ CAA ⎫ gln CAG ⎭	CGU ⎫ CGC ⎪ arg CGA ⎪ CGG ⎭	U C A G
	A	AUU ⎫ AUC ⎬ ileu AUA ⎭ AUG met	ACU ⎫ ACC ⎪ thr ACA ⎪ ACG ⎭	AAU ⎫ asn AAC ⎭ AAA ⎫ lys AAG ⎭	AGU ⎫ ser AGC ⎭ AGA ⎫ arg AGG ⎭	U C A G
	G	GUU ⎫ GUC ⎪ val GUA ⎪ GUG ⎭	GCU ⎫ GCC ⎪ ala GCA ⎪ GCG ⎭	GAU ⎫ asp GAC ⎭ GAA ⎫ glu GAG ⎭	GGU ⎫ GGC ⎪ gly GGA ⎪ GGG ⎭	U C A G

The abbreviated names of amino acids are as follows: **ala** = alanine, **arg** = arginine, **asn** = asparagine, **asp** = aspartic acid, **cys** = cysteine, **gln** = glutamine, **glu** = glutamic acid, **gly** = glycine, **his** = histidine, **ileu** = isoleucine, **leu** = leucine, **lys** = lysine, **met** = methionine, **phe** = phenylalanine, **pro** = proline, **ser** = serine, **thr** = threonine, **try** = tryptophan, **tyr** = tyrosine, **val** = valine.

Note: do not attempt to learn this table.

Note

(1) The code 'words' are groups of three bases, each group is called a codon.

(2) A codon does not code for more than one amino acid.

(3) One amino acid may be coded for by related codons. Very often, the first two letters (bases) are the same in all the codons for one amino acid, only the third letter differs (e.g., alanine is coded for by GCU, GCA, and GCG).

Another way of saying this is to say that for many amino acids only the first two letters (bases) of the codon are important, the third letter can be anything.

(4) Some amino acids are coded for by any one of up to six codons, e.g., serine and arginine; others by only one, e.g., methionine and tryptophan. The few codons that do not correspond to an amino acid have been found to be the punctuation sequences of the genetic code. For example, the sequences U–A–A, U–G–A, and U–A–G function as full stops. No amino acid is incorporated into the polypeptide chain where any of these codon triplets occur. Instead the polypeptide is released from the ribosome as a completed protein chain. In a similar way the triplet codon A–U–G marks the starting-point for the assembly of a polypeptide sequence. Thus all proteins originally start with methionine, but this may subsequently be removed by an enzyme.

It seems that a particular sequence of bases always stands for the same amino acids, in whatever organism it occurs. An mRNA sequence of UGU–UAU–CCU would be translated by human cells into the amino-acid sequence cys-tyr-pro, just as it would be by bacterial cells or by plant cells.

It should now be clear that any single cell can, using various sequences of codons, make any number of proteins, differing in content and sequence of the amino acids, and differing in length. Further, a different organism using the same code, can make its own unique set of proteins. In spite of the enormous variety of living organisms, it turns out that all those so far examined (*E. coli*, Guinea Pig, South African Clawed Toad, Yeast, Tobacco, Man) do, in fact, use the same code; i.e., the code is a universal one.

Having examined the manner in which proteins are synthesized, now let us return to the biology of whole cells. Growth and reproduction involve both synthesis of new material and division of cells. Division involves duplication of the cell's DNA, followed by its distribution to the daughter cells. If this distribution process is not carried out exactly, one of the daughter cells will lack some part of the DNA—i.e., the genetic information—and will fail to behave normally. Because DNA acts as a template in replication and transcription, it is essential that it should be copied exactly and distributed equally at cell division. Exact replication at the molecular level is made possible by the double helical structure of the molecule and the specific base pairing of its component parts, as we have already seen.

function of cell division

Now answer *SAQ* 25.

17.11 Cell Division and the Behaviour of Chromosomes

Study comment

> The manner in which DNA is distributed in dividing cells is considered by means of a structured exercise based on the behaviour of chromosomes in dividing cells. It is important to understand the main features of division and the genetic consequences of them. Do not attempt to memorize the details however.

Organisms grow over long periods of time. Microscopic examination of parts of their bodies at different times during growth shows that the number of cells increases. Cell multiplication is part of the process of growth. If this is not obvious, consider a seed and its component cells. During growth, the seed grows into a large plant which produces many new seeds. The first seed, and all the next generation of seeds, each contain similar numbers of cells, as can be shown by counting them under a microscope. In addition, there are cells of the vegetative part of the plant which were also derived from the cells of the original seed. Hence, cell multiplication occurs during the growth of the plant from the initial seed.

Root tips are convenient parts of plants for examining the process of cell multiplication. In roots, cell multiplication is localized, but not strictly confined to the tip of the root.

You are provided with a drawing based on a photograph of cells from the tip of the root of a *Crocus* (p. 39). From observations of this drawing, and your inferences about it, some of the features of cell multiplication can be examined and their implications exposed.

The biologists who worked out the details of cell multiplication were not able to view living cells down the microscope. They did not have suitable techniques for observing living cells. They could not see directly, therefore, the process of cell division, as you will see it on film in this Unit's television programme. Hence, they had to put together their own interpretations of a dynamic process from viewing dead cells in which the dynamic process had been arrested. This is fundamentally the exercise we want you to perform.

It is still a common problem in biology to have to interpret a dynamic process in living cells from images of dead cells. For example, although electron microscopy is so powerful a method for looking into cells, most of the techniques used with the electron microscope require the use of dead cells. Electron microscopy, therefore, presents the same problems of interpretation to modern biologists as light microscopy often presented to biologists fifty or more years ago.

In putting together a dynamic interpretation of the static diagrams, you have to consider several problems and answer many questions.

Question 13

It is first necessary to ask, 'What are the circumstances under which this living root tip was transformed into a static picture?' This is not just one question, it conceals many. Before going any further, list the circumstances or conditions that you would like to have information about relevant to this question, as a basis for proceeding further. Compare your list and that given on pp. 73–74.

The diagram shows a collection of cells which were all killed at the same time. Inspection shows that not all cells look similar. This leads to the possibility that these cells are at different stages of the multiplication process. If this is so, a comparison of the appearance of these cells should permit reconstruction of the overall process of cell multiplication. The situation is analogous to that of a still film. A single still photo of one person running would not allow us to reconstruct the whole running process. However, a single still photo, showing many people running out of step with each other, would allow us to reconstruct the movements of the body and limb during the running process.

Question 14

What would be the implications of a group of multiplying cells prepared in the same way as those in the diagram, if they all looked similar to each other?

Question 15

In such a situation, could you reconstruct the cell multiplication process from a single diagram?

Now check your answers with those on p. 74.

Examine the cell labelled 1 in the drawing and identify the cell wall, cytoplasm, nucleus, and the nuclear contents. Check your identification against that on p. 74.

The stain used in the preparation of material reacts strongly with DNA and much less strongly with other substances in the cell. Notice that in cell 1 there are no obvious thread-like structures inside the nucleus. The deeply stained granules are apparently arranged haphazardly in a roughly circular shape. Part of the preparative procedure used involves gently squashing the stained cells between a microscope slide and cover slip. This spreads the cell contents flat, distorting the real spatial relationships in the cell but making observations easier. The fact that the nuclear contents in cell 1, for instance, do not spread widely, suggests that the contents are prevented from spreading by some structure not visible in the drawing or photograph on which it is based. Observations of similar cells prepared using other stains show that a nuclear membrane is present around the nuclear contents of cells similar to cell 1.

Question 16

In which other cells is the presence of a nuclear membrane suggested by the shape of the heavily stained material of the nucleus?

Question 17

In which cells is an intact nuclear membrane unlikely to be present?

Examine cells 8 and 9. Identify the cell wall and the deeply stained thread-like chromosomes in both cells. Notice that the shape and size of the chromosomes differs in the two cells; and the position of the chromosomes in the cell, with respect to other chromosomes in the same cell, also differs. Check your answer against that on p. 75.

Question 18

In which three major ways do all the chromosomes in cell 8 differ from those in cell 9?

Question 19

In what way does the arrangement in space of all the chromosomes in cell 8 differ from that in cell 9?

Now check your answers with those on p. 75.

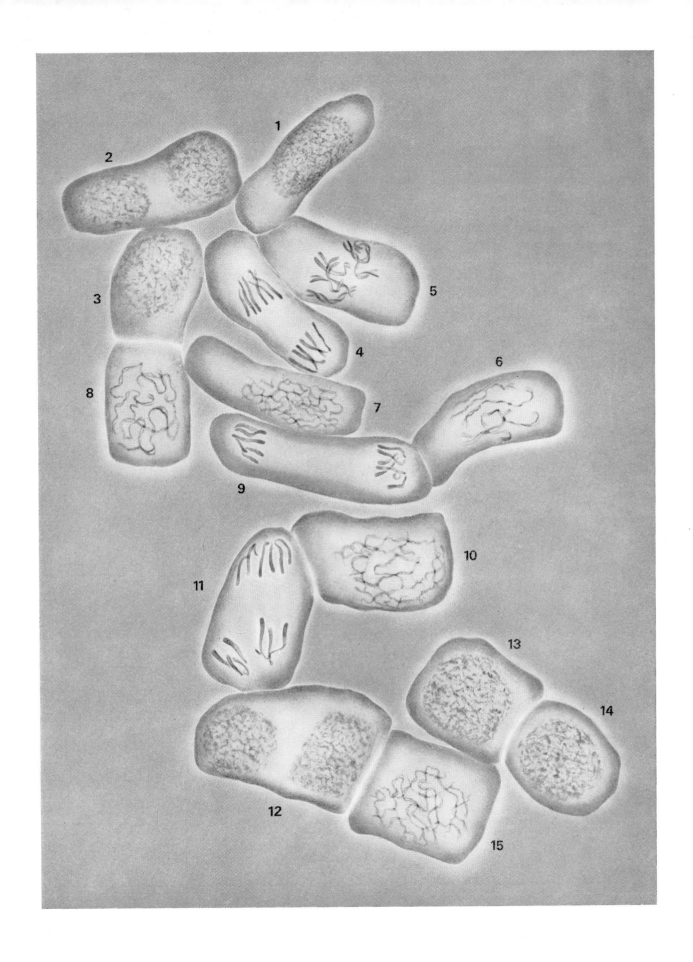

Drawing based on a photograph of cells from the tip of a root of a Crocus.

Classify each cell in turn according to the scheme given opposite. For each cell in turn, start with the question in the upper left-hand corner and then follow the line to the appropriate answer you have given to the first question and every other subsequent question. Write down the number of every cell in its box on the right-hand side appropriate to its characteristics.

If the sample of cells, shown in the photograph, comes from a population of cells, all dividing asynchronously (out of step with each other), then the sequence of events can be reconstructed if we assume that enough of the stages in the multiplication process are seen in the sample. In the region of the *Crocus* root, from which this preparation was made, cell division is going on continuously. Cells divide to produce more daughter cells and so on. Between each division, the cells grow until they reach their mature size and then they divide. Because division is continuous, it cannot be said to have a start and a finish. However, for our purposes, we can reconstruct the events in one division cycle, starting with one cell and ending with two daughters.

Question 20

Ring the letter corresponding to the group of cells that most nearly represents the start of a division cycle, Group A, B, C, D, E.

Question 21

Ring the letter corresponding to the group of cells that, in your opinion, most nearly represents the end of a single division cycle, Group A, B, C, D, E.

Question 22

Underline the sequence that best represents the series of events between the start and the end of a division cycle:

$$\text{Start} \rightarrow C \rightarrow D \rightarrow E \rightarrow \text{Finish}$$
$$\text{Start} \rightarrow D \rightarrow E \rightarrow C \rightarrow \text{Finish}$$
$$\text{Start} \rightarrow E \rightarrow C \rightarrow D \rightarrow \text{Finish}$$
$$\text{Start} \rightarrow C \rightarrow E \rightarrow D \rightarrow \text{Finish}$$
$$\text{Start} \rightarrow E \rightarrow D \rightarrow C \rightarrow \text{Finish}$$
$$\text{Start} \rightarrow D \rightarrow C \rightarrow E \rightarrow \text{Finish}$$

Now check your answers with those on p. 76.

Question 23

If the order of events in cell multiplication is $----\rightarrow B ----\rightarrow A$, what must have happened to each *chromosome* in one of the nuclei between stage B of one cycle and stage D of the next division cycle of this nucleus?

Question 24

Stage E is followed by stage C. Between these two stages, each chromosome separates into two parts and the two parts move away from each other to opposite ends of the cell. If every chromosome behaves in this way, what does this suggest about the inherited characteristics of the two daughter cells when compared to their parent cells?

All sexually reproducing organisms start life as a single cell—the fertilized *zygote*. This zygote divides by the processes we are examining. Its daughters divide, and their daughters divide, eventually building up a large cell population that becomes a new organism.

START

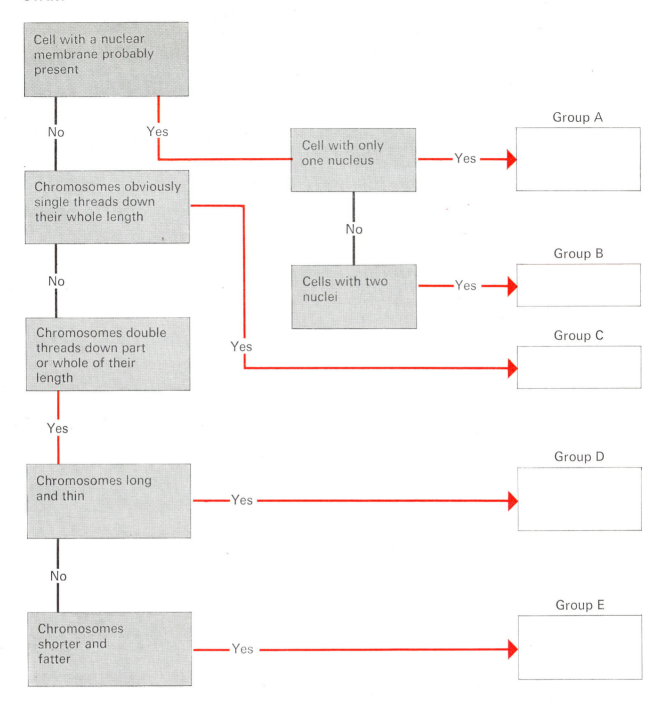

Check the result of your classification with that on pp. 75 and 76

Question 25

What can you conclude about the similarity or otherwise of the inherited characteristics of the daughter cells, compared with that of the zygote?

Check your answers against those on p. 77.

There are then two principle 'events' during mitosis that are of great importance:
  (1) replication of the chromosomes;
  (2) exact distribution of the chromosomes to the daughter cells.

As a direct consequence of these, the conclusion that living organisms are united by genetic continuity follows. Every chromosome ought to be an exact copy of a pre-existing chromosome and this line of descent can be projected into the distant past—perhaps even to the time when life originated.

The idea of genetic unity is considered further in section 17.14 of this Unit. In the case of organisms that reproduce asexually, that is without gametes and zygotes, it is even easier to see that the consequence of duplication of chromosomes, followed by their exact and orderly separation in mitosis, constitutes a genetic link between parents and offspring that must have continued since the first cells became organized.

genetic unity

This genetic continuity based on the mitotic process does not of course explain the variation and differences between living organisms. This can, however, be accounted for by a second type of nuclear process called meiosis which will be examined in Unit 19.

42

17.12 DNA, Genes, Genotype and Phenotype

Study comment

> The characteristics of cells of organisms are related to the environment and to the genetic information inherited from previous generations. This important relationship is discussed in outline.

According to the analysis we have been making so far, the unit of the genetic code could be defined as a single codon, three bases in length, but it could also be defined as a single base pair or as the whole sequence encoding for a whole polypeptide chain.

It we look at this in another way, new terms have to be used. One way of finding out about inheritance is to breed organisms and compare parents with offspring. This type of investigation has led to a system in which the inheritance of characters can be described in terms of the inheritance of units called genes. A gene is identified as the unit that affects, or appears to affect, a distinct character of the whole organism. Not only are genes fairly large and complex units in chemical terms but, because visible characters are usually the end-product of a series of chemical reactions, it is difficult to define a gene in such a way that the definition fits every case. The most general description is that one gene affects one or more distinct characteristics, for example, the colour of the eyes in a fruit-fly.

Since they are complex chemical units, genes may have a number of different forms. In this case, one form of the eye-colour gene prevents pigment formation altogether, so the eyes are white; another form has a less extreme effect, and the eyes are pink; and another a less extreme effect still, and the eyes are an apricot colour. The 'normal' form, that most usually found, determines red eyes. However, they are all different forms of the same gene.

The colour of the eyes does not depend, however, solely on this particular gene in any of its forms, but also upon the activities of a number of other genes.

It has been satisfactorily established now that the unit of heredity which a geneticist identifies in this way as a gene is, in fact, that length of DNA which determines the structure of one particular protein chain. Usually the characteristics which geneticists look at depend upon the activity of that protein.

The total number and kinds of genes encoded into the DNA of a nucleus is described as *the genotype* of the nucleus. But, unless the genes can be described in terms of the sequence of bases of which they are made up, some other system of description must be used.

genotype

While the genotype of a cell or organism refers to the genetic information encoded in the nuclei, a description of an organism could be written based on its visible distinct characters. Such a description is called the *phenotype* of the organism. For example. the height of a man is part of his phenotype. It is partly determined by his genotype, but we can describe its effect, that is on his height. Depending on the refinement of our techniques of observation, the phenotypic effect of a gene can be described in greater or less detail. For example, Yanofsky obtained a strain of bacteria that was first identified from colonies which failed to grow on a food medium lacking the amino acid tryptophan. The phenotype 'failure to grow' is readily

phenotype

detected by the naked eye if the proper growth media are used. Yanofsky, a biochemist, was also able to show that this strain of bacteria lacked a particular enzyme, and so lacked the ability to convert indole and serine into tryptophan. The phenotype is observed and described in a different way, but the genotype is the same. Ultimately, Yanofsky was able to describe the phenotype of a particular strain in terms of the presence or absence of one amino acid at a certain position in a protein, the enzyme tryptophan synthetase. Alteration of this particular amino acid inactivates the enzyme. The one genotype can be defined as three different phenotypes in three phenotypic systems. This can be represented diagrammatically as:

> ——→PHENOTYPE I —fails to grow on a normal food
> medium
>
> GENOTYPE——→PHENOTYPE II —fails to synthesize the active enzyme
> trypophan synthetase
>
> ——→PHENOTYPE III—one particular amino acid substitu-
> ted in the normal sequence

Incidentally, there is a philosophical point here: the lower the hierarchical level at which the phenotype is described, the closer the description comes to describing the genotype (see Unit 16 for a discussion of hierarchical levels). At the level of describing the DNA sequence, genotype and phenotype are indistinguishable and so the idea of a genotype is purely notional.

Both genotype and phenotype are words which can be used to refer either to a single gene or characteristic or to several genes and the characteristics they determine, or to the whole composition of an organism, depending on how much one wishes to discuss or describe in any particular context.

In our bodies, for instance, cartilage cells not only produce cartilage—the slippery tough substance that covers the ends of the bones—but also divide and produce a new generation of cartilage cells. In terms of protein synthesis, we might regard the daughter cells as receiving from their parents the kind of genetic information to ensure that the cell produces its own kind of chemicals during its lifetime.

17.13 Differentiation of Cells

Study comment

> **Different types of cells may be produced from a single parent cell, this process is differentiation. In this section, differentiation is discussed and a hypothesis is put forward to account for some features of the process.**

Not all cells behave in the straightforward way described in the previous section. Consider for instance a human egg cell. On fertilization, this cell, now a zygote, divides repeatedly and grows rapidly. The products of division grow, divide, and develop into a very wide variety of different types of cells. Each different type of cell has a different set of proteins. The proteins it makes determine its shape, size, rate of division and metabolic activity. However, since each of these new kinds of cells—muscle, blood, nerve, and liver cells—is a descendant of the original zygote, it is evident that the single zygote's DNA must have contained all the genetic information necessary for the whole operation of every type of cell of the normal adult body. But the individual types of cell behave as if they had only information relating to their particular activity. This effect could be achieved in one of several ways. Either the genetic information in the zygote could be unequally distributed, so that its descendants received only their own set of information. Or the genetic instructions could be equally distributed, but used differently by different descendants.

> **Use each of these hypotheses in turn to predict what would happen to the two daughter cells produced by one division of a human zygote, if they were to fall apart from each other and continue to develop.**

What actually happens is that both daughter cells grow, develop, divide, and produce two new organisms—identical twins. Identical because they share a common store of genetic information. So the idea of unequal distribution at each division does not appear to be supported. Similar failure of this hypothesis occurred when isolated plant cells were shown to be able to develop into whole plants. This investigation and its results are summarized in Figure 8.

Here it looks as though the starting cell, although it normally behaves as if it did NOT contain a complete set of genetic instructions, can, under the special conditions of this experiment, be made to grow and develop into a whole plant able to produce flowers and seeds. This shows that it does, in fact, contain the whole of the genetic instructions appropriate to that whole organism.

Evidently the carrot cells are able to translate and transcribe the complete genetic instructions, although normally they do not do so. Normally they translate and transcribe only a small part of the whole information present in their nuclei. It is as if the transcription/translation process can be switched on and off. When switched on, selected parts of the information encoded in the DNA in the nucleus are transcribed and translated into proteins in the cell. It is these that define the cell as being of a particular type when mature. So, according to this hypothesis, a muscle cell, for example, becomes different from other cells as the result of the appropriate inherited information in the nucleus being switched on, resulting in the formation of proteins in the cell appropriate for its function, with the consequent changes in visible characteristics and behaviour of the cell.

switch-on/switch-off hypothesis

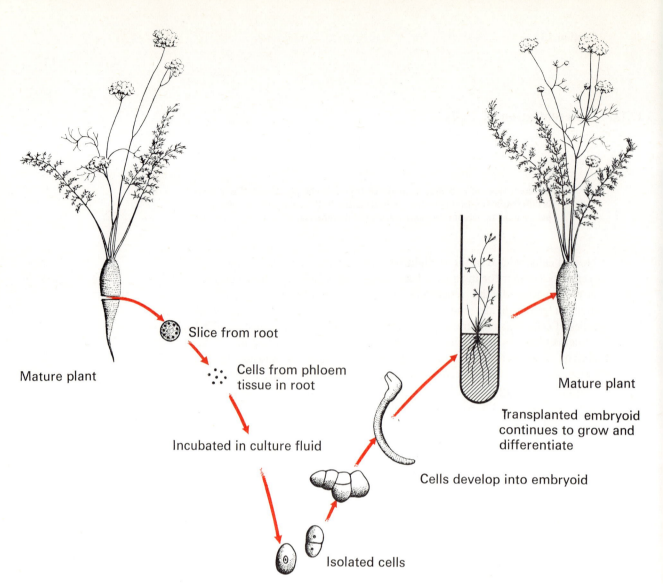

Slice from root

Cells from phloem
tissue in root

Mature plant

Incubated in culture fluid

Isolated cells

Cells develop into embryoid

Transplanted embryoid
continues to grow and
differentiate

Mature plant

Figure 8

*Diagram of sequence of events in regeneration investigations using isolated carrot cells.
Modified from Ebert 1965.*

Support for the switch-on/switch-off hypothesis has been obtained experimentally from the study of embryo development in the African Clawed
Toad (*Xenopus*) by Gurdon. He used the technique of nuclear transplanta-
tion, taking nuclei from mature cells lining the intestine of a toad and
inserting them into zygotes from which the nuclei had been removed. If the
nuclei from intestinal cells had switched off once they had completed their
differentiation, then the zygotes into which they were inserted ought not to
develop into normal tadpoles. But some of Gurdon's composite zygotes
did develop normally, so clearly the nuclei of this animal were stable and
had not altered; that is, the nuclei always retained all their instructions
in a usable form, in spite of the fact that the cells from which the nuclei
were derived had completed their development.

Other work has shown that, in spite of this stability, the translation of
genetic instructions into a fully differentiated organism can be upset.
Sometimes this is done accidentally, as in the thalidomide tragedies of
recent years. Some pregnant women who had been taking the drug
thalidomide later produced deformed babies. Regrettably more common,
but no less serious, is the interference which can result when a pregnant
woman becomes infected with Rubella virus, which causes German measles.
In this case, differentiation and development of the unborn child is abnor-
mal. Exactly how thalidomide and the Rubella virus affect the unborn
infant is not known. A large number of chemicals that can produce such

deformities are known. These chemicals are called teratogenic agents. The possibility of teratogenic effects is one of the hazards that must be borne in mind when any new chemical is introduced into the environment as a food additive, or as a drug, or for any other purpose.

Growth, development, and differentiation during the early life of all organisms must be an ordered and controlled process (see Unit 16 for other aspects of control inside cells). The events must occur in the correct sequence, and end at the proper time. Even in adult, fully differentiated organisms, wound healing and tissue repairs (even to the extent of re-growth of missing parts in some organisms) indicate that the cellular events of chemical synthesis can go on in a highly ordered fashion. Essentially, these sequential events must result from control over the transcription and translation processes inside the living cells, plus the regulated switching-on and switching-off of particular portions of the information encoded in the cells, DNA or the transcribed RNA. Although the insight we have at the moment into these controlling processes is small, the great success that has rewarded investigators working in the field of molecular biology of inheritance and protein synthesis encourages belief that an accurate description of the controlling process will be written. This, however, is in the future. Here and now, we can at least understand how genetic instructions are encoded, and the code itself has been decoded. At the same time, the discovery that the genetic code, together with the transcription and translation processes, appear to be universal—the same in all organisms so far examined—is a tremendous intellectual landmark in our attempt to understand ourselves and the world in which we live.

17.14 Genetic Continuity as a Consequence of DNA Replication and Mitosis

Study comment

> The consequences of the previous sections of the Unit are explored and some of the implications pointed out.

It is easy to see how beautifully the helical DNA molecule is adapted to its informational function, while at the same time performing in a self-replicating way. This, however, should not be allowed to detract from a proper sense of wonder and an emotional response to the exquisite precision with which structure and function are related. The apparent simplicity of the code of four different bases, the exactness of the transcription of the genetic language, and the economy of the translation process are as aesthetically satisfactory as a great painting, music, an elegant mathematical equation, or a beautiful human being. In addition the sure knowledge that all organisms are essentially alike as far as their genetic apparatus is concerned allows us to see that all organisms are united as a result of their common origin and also as a consequence of DNA replication. The similarities and differences alike result from processes of cell division that essentially replicate pre-existing spatial patterns of nucleotides in DNA molecules. Differences arise when the process of replication fails in exactness. The alteration of amino-acid composition in proteins, or even an alteration of the order in which the same amino acids are arranged in the polypeptide chains, results from an alteration in the replication of pre-existing DNA. How such alterations occur, and the effects they can have form the subject of Unit 19. At this point, however, it is a little unnerving to discover that our differences from each other and the similarities to each other depend to a great extent on the similarities in the DNA molecules in all our cells, and then to go on from this point to realize that our DNA—humanity's DNA—has resulted from replication of previous DNA at each cell division, generation after generation as far back in time as life itself. And how that all started will be discussed in Unit 21.

Now answer *SAQs* 27–40.

Parallel and Background reading

Chapters appropriate to Unit 17 are given in brackets after each title.

M. J. Berrill, *Biology in Action*, Heinemann, 1967 (Chapters 6 and 7).

P. B. Weisz, *et al.*, *The Science of Biology*, McGraw-Hill, 3rd edition, 1967 (Chapters 20, 29).
(Laboratory and Study Guide not required.)

S. D. Gerking, *Biological Systems*, Saunders, 1969 (Chapters 2, 22).

A. G. Loewy and P. Siekewitz, *Cell Structure and Functions*, Holt, Rinehart and Winston, 2nd edition, 1969 (Chapters 8, 14, 15).

J. A. Ramsay, *The Experimental Basis of Modern Biology*, CUP, 1965 (Chapters 22, 23, 24)

H. R. Mahler and E. V. Cordes, *Basic Biological Chemistry*, Harper and Row, 1968 (Chapters 5, 20).

J. D. Watson, *The Double Helix*, Weidenfeld and Nicholson, 1968. Penguin, 1970.

Breaking the DNA Code

The experiments which led to the breaking of the code were in principle very simple. Given the sequence DNA → RNA → proteins, decoding the code involved: putting into a protein-synthesizing system DNA or RNA in which the sequence of bases was known; allowing the system to make protein; harvesting this protein; and then analysing it to find out what amino acids it contained and in which order they had been linked together. In practice, extraction and preparation of the necessary bits and pieces from cells was difficult. However, the practical difficulties were overcome; RNA of known composition was added to a mixture of amino acids, ribosomes, enzymes, and ATP in test-tubes, and the polypeptides made in the test-tubes were examined. In this way Nirenberg and Matthaei in 1961 found that synthetic RNA they made in which all the bases were uracil led to the production of polypeptides containing one amino acid only—phenylalanine. So the codon for phenylalanine must have been U–U–U.

The Nirenberg and Matthaei technique was extended to the production of RNA containing a mixture of two or even three bases. But here interpretation of the results is more difficult. The difficulty stems from the method by which the RNA is prepared. To do this, a solution containing the required bases is mixed with a solution containing an enzyme that links the bases in sequence. The order in which the bases are linked is a random one. In the early experiments, this did not matter, because only one kind of base was used. But where more than one base is used, there are more ways than one in which they can be linked together. For example, if equal amounts of uracil and adenine are used, then eight sequences are possible: U–U–U and A–A–A; U–A–A and A–A–U; U–U–A and A–U–U; U–A–U and A–U–A. Each of these sequences would be expected in equal amounts. So polypeptides with eight amino acids in equal amounts would be made. It would, however, still not be possible to determine which RNA sequence corresponded with which amino acid.

The way round part of this difficulty is neat and elegant. Instead of mixing bases in equal amounts, mix them in unequal amounts. Then the various combinations of bases would be made in unequal amounts and the amino acids would be incorporated in unequal amounts. If, however, the amount of any particular amino acid incorporated corresponded to the amount of a particular RNA sequence, then the RNA triplet corresponding to that particular amino acid would be identified.

For example, if in the original mixture there was five times as much uracil as there was cytosine, then if codons were assembled in random order, the following sequences would be expected: U–U–U and C–C–C; U–C–U and C–U–C; U–U–C and U–C–C and C–U–U. But, because more uracil than cytosine was in the original mixture, then sequences containing more uracil than cytosine would be formed more frequently than sequences containing more cytosine than uracil. To be specific, given five uracils to each cytosine, the chance of the first base in a codon being uracil is $(5/6)$, similarly the chance of the second base being uracil is $(5/6)$ and of the third base being uracil is $(5/6)$. So the chances of the U–U–U codon being formed is $(5/6) \times (5/6) \times (5/6) = 125/216$. The chance of the codon

C–C–C being formed is then $(1/6) \times (1/6) \times (1/6) = 1/216$. By a similar argument the chances of sequences containing 2 uracils and 1 cytosine being formed is:

U–U–C	$(5/6) \times (5/6) \times (1/6) = 25/216$
U–C–U	$(5/6) \times (1/6) \times (5/6) = 25/216$
C–U–U	$(1/6) \times (5/6) \times (5/6) = 25/216$

and of the sequences of 1 uracil and 2 cytosines is:

U–C–C
C–U–C } 5/216 in each case.
C–C–U

Now, if the proportions of the amino acids incorporated are compared with the predicted frequencies of the various RNA sequences, it will be possible to pair up sequences with amino acids.

There is one class of RNA codons which cannot be decoded by either of these methods. If a mixture of adenine and cytosine is made containing twice as much adenine as cytosine, then equal amounts of the three codons A–A–C, C–A–A, and A–C–A would be formed. It is now known that the amino acids which correspond to these codons are asparagine, glutamic acid, and threonine. But which codon corresponds to which amino acid could not be discovered by this method. In order to determine the code words exactly, that is to determine both the bases they contain and the sequence of the bases in them, another technique was devised by Nirenberg and Leder in 1964.

When proteins are being synthesized, two different RNA structures have to be 'bound' to the ribosomes. One is the messenger RNA, and the other is an amino-acyl transfer RNA, which 'recognizes' a 'triplet' on the bound messenger.

Nirenberg and Leder discovered that amino-acyl-transfer RNA could be bound to ribosomes by extremely short lengths of messenger RNA. All that was necessary was a messenger RNA of three nucleotides, a 'trinucleotide'. Of course, no peptide synthesis occurs, only the binding. However, the particular amino-acyl-transfer RNA that is bound depends upon the trinucleotide. For example, if a mixture of amino-acyl-transfer RNA, ribosomes, and the trinucleotide UUC is made, the ribosomes have only phenylalanyl-transfer RNA bound to them. If the mixture contains instead the trinucleotide UCU, then only seryl-transfer RNA is bound.

The ribosomes, and what is bound to them, can be separated from the rest of the mixture by filtration. Nirenberg and Leder suggested that these trinucleotides are the 'codons' for the amino acids they bind to ribosomes through tRNA molecules, UUC being a codon for phenylalamine, UCU for serine. The code they worked out from these experiments agrees completely with the 'unordered' code determined by the use of artificial random messengers, but is, of course, much more complete. It was found that nearly every one of the 64 possible triplet codons specified one particular amino acid. Only three codons appeared to be meaningless, i.e., failed to bind any amino-acyl-transfer RNA. They were UAA, UAG, and UGA. These have come to be known as 'nonsense' codons, but it is clear that at least one of them, UAA, is not really nonsense, but is actually a punctuation mark, indicating the point at which amino acids stop being joined together and the protein chain determined by a particular messenger is completed. More recent work shows that UAG and UGA are also punctuation codons.

Now answer *SAQ* 26.

(Where a question is not explicitly asked, you are expected to select the correct statement(s) from the set given.)

Section 17.2

(*Objectives 1a, 1d, 1i, 1j, 1k*)

The diagram below is of a section of DNA during replication. Refer to it to answer the questions which follow.

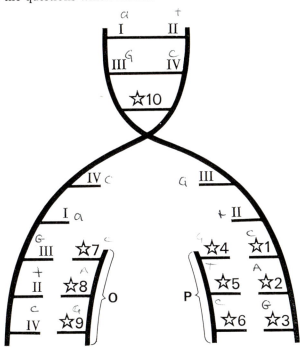

Question 1

At locations marked I, II, III, and IV there would be:

A   ATP

B   RNA

C   sugars

D   phosphates

E   bases

Question 2

If I is adenine then II is:

A   thymine

B   cytosine

C   guanine

D   a purine

Question 3

If III is guanine then IV is:

A   thymine
B   cytosine
C   a purine
D   adenine

Question 4

At location ☆10 we find:

A   I and II
B   III and IV
C   II and I
D   IV and V
E   cannot be predicted

Question 5

At location ☆1 must be:

A   I
B   II
C   III
D   IV
E   unknown

Question 6

At location ☆2 must be:

A   I
B   II
C   III
D   IV
E   unknown

Question 7

At location ☆3 must be:

A   I
B   II
C   III
D   IV
E   unknown

Question 8

On O and P adenine will appear in position:

A   ☆4
B   ☆5
C   ☆6
D   ☆7
E   ☆8

Question 9

Thymine will appear on O and P in position:

A   ☆5
B   ☆6
C   ☆8
D   ☆5 and ☆8
E   ☆6 and ☆8

Question 10

Cytosine will appear on O and P in position:

A   ☆4 only
B   ☆6 only
C   ☆7 only
D   ☆4 and ☆8
E   ☆6 and ☆7

Question 11

Guanine will be present in positions:

A   ☆1 and ☆6
B   ☆2 and ☆4 and ☆7
C   ☆3 and ☆4 and ☆9
D   ☆9 only
E   none of above

(*Objectives 1a, 1c, 1k*)

The following questions refer to the partially labelled diagram, which illustrates a segment of a particular giant molecule produced in living cells. Examine the diagram then answer the questions.

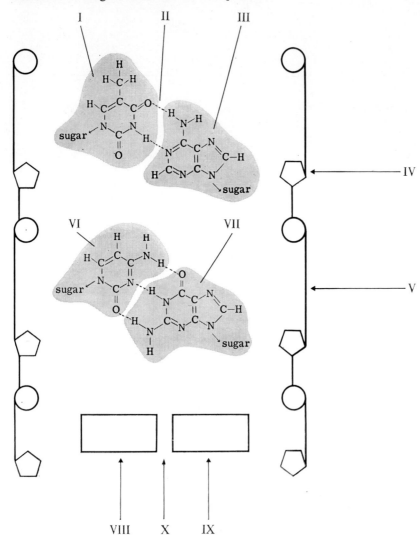

Question 12

The diagram illustrates a segment of a

A   DNA molecule

B   tRNA molecule

C   mRNA molecule

D   polypeptide molecule

Question 13

The structure represented by I in the diagram is:

A   a base; a pyrimidine; thymine

B   a sugar; ribose; thymine

C   an enzyme; a lipid; an organelle
D   a peptide; an amino acid; thymine
E   a base; a purine; adenosine

Question 14

Label II represents a:
A   hydrogen bond
B   polymerase
C   peptide bond

Question 15

Structure III represents:
A   deoxyribose
B   ribose
C   adenine
D   guanine
E   cytosine

Question 16

Structure IV represents:
A   a base; a pyrimidine; cytosine
B   an amino acid
C   a phosphate
D   a sugar; deoxyribose
E   ribose

Question 17

Structure V is a bond between:
A   a purine and a pyrimidine
B   a sugar and a base
C   a phosphate and a base
D   a sugar and a phosphate

Question 18

The structures labelled VI and VII are, respectively:
A   a pyrimidine and a purine
B   a purine and a pyrimidine
C   ribose and deoxyribose
D   an enzyme and a substrate
E   a substrate and an enzyme

Question 19

Which of the following are 'complementary' to each other?

A   the sugar and the phosphate

B   thymine and cytosine

C   the members of a base pair

D   adenine and guanine

E   deoxyribose and ribose

Question 20

According to the base-pairing theory, if the structure labelled VIII were adenine, then IX would be:

A   a purine

B   ribose

C   deoxyribose

D   phosphate

E   thymine

Question 21

If VIII were guanine, then X would represent:

A   two covalent bonds

B   two peptide bonds

C   three hydrogen bonds

D   two hydrogen bonds

Question 22

In the molecule, of which only a segment is represented in the diagram, the number of adenine molecules equals the number of (————) molecules.

A   cytosine

B   phosphate

C   deoxyribose

D   thymine

E   guanine

Question 23

In the molecule diagrammed, the number of purine molecules equals the number of (————) molecules.

A   phosphate

B   uracil

C   ribose

D   deoxyribose

E   pyrimidine

Question 24

Which of the following is true of uracil?

A   It is present in RNA but not DNA, and is a pyrimidine complementary to adenine.

B   It is present in messenger RNA but not in tRNA, and is a pyrimidine complementary to cytosine.

C   It is present in tRNA but not messenger RNA, and is a purine complementary to guanine.

D   It is present in ribosomal RNA but not messenger RNA, and is a purine complementary to thymine.

E   It is present in messenger and tRNA but not ribosomal RNA, and is a pyrimidine complementary to guanine.

Section 17.10.1

Question 25

(*Objectives 1e, 1f, 1g*)

For the DNA segment below write down:

(i) the mRNA segment you would expect to be made corresponding to it;

(ii) the tRNA anticodons you would expect to 'read' each mRNA segment and;

(iii) using Table 3 on p. 35, section 17.10.1, the sequence of amino acids of the polypeptide into which you would expect the DNA segment to be translated. Work first from left to right and then make another translation working from right to left.

AU C AC G UU UC GAG U CAAGAA UU GC AAC G UA

DNA segment: T A G T G C A A A G C T C A G T T C T T A A C G T T G C A T

A UC A CG UUUCG AG UCAAG AAUUGC AACGUA

Appendix 1

Question 26

(*Objective 1g*)

Suppose that a synthetic RNA was made from a solution containing 80 per cent adenine and 20 per cent uracil. The proteins produced in a cell-free system under the direction of the synthetic RNA were found to contain amino acids in the following proportions:

4 times as many isoleucines as tyrosines;

16 times as many isoleucines as phenylalinines;

16 times as many lysines as tyrosines.

What triplet codons are probably specifying each of the amino acids mentioned?

Self-Assessment Questions 27–40

(To be attempted after the Unit has been completed.)

Section 17.1

Question 27

(*Objective 3*)

The following diagrams represent models that describe events in the life cycle of cells. X is the mother cell, and Y_1 and Y_2 are the daughter cells. If *a* represents raw materials, *b* represents increase in cell size and *c* represents cell division, then which diagram best shows the relationship found in most cells?

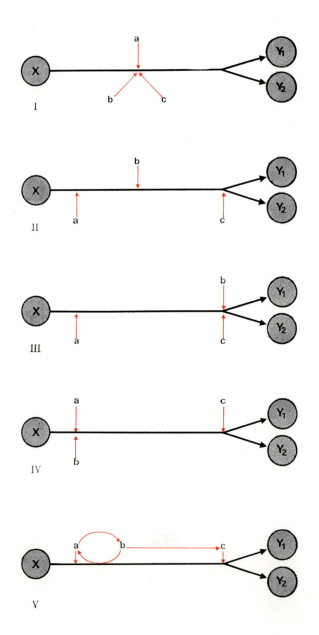

Sections 17.3 and 17.6

Question 28

(*Objectives 1b, 1c*)

Both tRNA and DNA are believed to be characterized by which of the following?

A   Both are composed of purine, pyrimidine, sugar, and phosphate groups.

B   Both are unbranched linear macromolecules.

C   Both of the above.

D   None of the first two above.

Sections 17.9 and 17.10

Question 29

(*Objective 3*)

The reactions catalysed by RNA polymerase and DNA polymerase are similar in:

A   that a large molecular weight product is formed

B   that for each mole of nucleotide incorporated into polymer, one mole of pyrophosphate is produced

C   all of the above respects

D   none of the first three respects listed above

Question 30

(*Objective 1g*)

A chemical derivative of the natural amino acid, l-phenylalanine, can be prepared in the laboratory. The false amino acid, p-fluorophenylalanine, when added to a growing culture of normal *E. coli* cells in minimal medium, gets incorporated into newly made protein. From what is known about the mechanism of protein synthesis, it would be reasonable to suspect that:

A   normal *E. coli* can admit the false amino acid into the cell

B   normal *E. coli* cells probably contain at least one enzyme that can 'activate' the false amino acid

C   normal *E. coli* cells probably possess at least one tRNA species that can accept the false amino acid and adapt it to some one or more codons

D   all of the above occur

E   none of the first three above occur

Section 17.9

Question 31

(*Objective 1g*)

From the following list, select the one factor which seems most important in assuring that amino acid X will be placed where the mRNA directs it to be placed.

A   ATP

B   the tRNA for X

C   the ribosome making the protein

D   DNA polymerase

E   the concentration of X in the cell

Section 17.10

Question 32

(*Objective 1f*)

Assume that poly UG contains codons for alanine but not for tyrosine, and that poly UAC contains codons for tyrosine but not for alanine. Under otherwise appropriate conditions for polypeptide synthesis, which of the following pairs of reagents will lead to *some kind* of polypeptide?

A   Poly UG + tyrosine + $sRNA_{tyr}$

B   Poly UAC + tyrosine + $sRNA_{ala}$

C   Poly UG + alanine + $sRNA_{tyr}$

D   More than one of the above

E   None of the first three above

The following questions (33–40) relate to Units 14, 15, 16, and 17.

Question 33

(*Objective 1i*)

In order for a chemical to serve as a code for heredity, it is essential that the chemical be:

A   able to form itself into long spiral chains

B   subject to replication

C   composed of pyrimidines and purines

D   easily changed

E   transferable to different strains of organisms

Question 34

(*Objective k*)

The principal nucleic acid of which chromosomes are composed is:

A   DNA

B   ribosomal RNA

C   transfer RNA

D   messenger RNA

E   uracil

Question 35

(*Objectives 1g and 3*)

The primary structure of a polypeptide is defined by a specific sequence of amino acids linked together by peptide bonds. In general, which of the following is true of primary structure?

A   It can be determined simply by the use of a device known as an amino-acid analyser.

B   It is unknowable by present-day analytical methods.

C   It can be ascertained today, but only by methods which include hydrolysis of the polypeptide to various degrees, analysis of the composition of the peptides formed, and identification of the N-terminal groups of these peptides.

D   It is of little interest to enzymologists.

E   None of the above.

Question 36

(*Objective 1g*)

If it is assumed that the primary action of DNA is to produce proteins, and that our current concepts of how proteins are produced are correct, which of the following cellular components need **NOT** be present in a zygote?

A   DNA.

B   Messenger RNA.

C   Ribosomes.

D   ATP, or a means of producing it.

E   Enzymes.

Question 37

(*Objective 3*)

If an *E. coli* cell were to be grown in a medium containing $(N^{15}H_4)_2SO_4$ as the sole source of nitrogen, then N^{15} could be reasonably expected in:

A the ribosomes

B the DNA

C the chromosomes

D the enzymatic proteins

E all of the above

Question 38

(*Objective 1g*)

If mammalian cells utilize leucine exclusively for protein synthesis, then feeding radioactive leucine to a growing cell culture will eventually label:

A the ribosomes

B the mitochondria

C the mRNA

D more than one of the above

E none of the first three above

Question 39

(*Objective 3*)

That the synthesis of DNA in cells is actually catalysed by the enzyme DNA polymerase has been made *virtually certain* by which of the following findings?

A A variety of this type of cell which lacks this enzyme must be supplied with DNA in its growth medium.

B The product made in a test-tube has approximately the same molecular weight as natural DNA.

C The product made in a test-tube always has the same biological properties (ability to direct enzyme synthesis).

D All of the above.

E None of the first three above.

Question 40

(*Objective 3*)

Which does not demonstrate the interrelationship between form and function?

A Spherical shape of the nucleus and storage of DNA.

B Infolding of inner mitochondrial membrane and arrangement of enzymes.

C Chloroplast lamellae and photosynthesis.

D Muscle fibrils and contraction.

E Extension of nerve cell and impulse conduction.

Self-assessment Answers and Comments

Question 1   E

Question 2   A

Question 3   B

Question 4   E

Question 5   D

Question 6   A

Question 7   C

Question 8   E

Question 9   A

Question 10   E

Question 11   C

Question 12

A a segment of a DNA molecule is correct.

B is incorrect because tRNA is not double-stranded, and has a clover-leaf shape. See text p. 31.

C is incorrect because mRNA is single-stranded along its whole length. See text p. 24.

D is incorrect because, if the segment were of a polypeptide, it would be single-stranded not double-stranded and all the molecular subunits would be amino acids, not 4 different kinds of subunit as shown in the diagram in the question.

Question 13

A is correct.

B must be incorrect, because thymine is not a sugar as stated in the item.

C must be incorrect, because an enzyme is not a lipid, and neither an enzyme nor a lipid is an organelle.

D must be incorrect because thymine is not an amino acid, nor is an amino acid a peptide.

63

E is incorrect because while adenine is a purine and purines are a class of base, all purines are double-ring chemical structures and not single-ring structures. See text on p. 12.

Question 14

A is correct.

B is not correct. Polymerase is the name of a particular kind of enzyme.

C is incorrect because a peptide bond involves formation of a band between NH_2 group of one amino acid and the COOH group of another. See text of Unit 13, p. 33.

Question 15

A is incorrect
B is incorrect $\Big\}$ for the structural formula of these see text p. 24.

C is correct, because adenine, which is a purine, is a single-ring substance and pairs in the DNA molecule with thymine—and structure I is thymine in this diagram.

D
E $\Big\}$ Both are incorrect because neither pairs with thymine and also III must be a purine not a pyrimidine so cytosine is impossible here. See text pp. 12 and 15.

Question 16

D is correct.

A is incorrect, because all pyrimidine molecules are single-ring compounds containing N atoms, together with C atoms in a six-membered ring. Structure V is a five-membered ring with no N atoms. See text p. 11.

B is incorrect because all amino acids contain a double-banded O atom which structure V lacks.

C is incorrect because phosphate groups are not ring compounds. See text p. 11.

E is incorrect, because the diagram represents DNA, and DNA contains deoxyribose not ribose. See text p. 24.

Question 17

D is correct.

A and B must both be incorrect, because neither structure IV nor structure V is a purine base or pyrimidine base.

C is incorrect, because the DNA bases are not bonded directly to phosphates.

Question 18

A is correct.

B is incorrect because VI is a single substance and so could not be a purine, while VII could not be a pyrimidine, because these are all single-ring structures.

C is incorrect, because both ribose and deoxyribose are five-membered single-ring structures.

D and E are both incorrect, because enzymes are not involved in this molecule.

Question 19

Complementary, in this context, means that the presence of one member implies the simultaneous presence of the other. So C is correct.

A is incorrect, because, although sugar and phosphate are bound in the molecule, they are not complementary.

B is incorrect, as both thymine and cytosine are pyrimidines and do not form a pair in the DNA molecule.

D is incorrect, as both adenine and guanine being purines do not form a pair.

E is incorrect, because deoxyribose and ribose do not occur together in DNA.

Question 20

E is correct.

A is wrong, because adenine is a purine and a base pair consists of a purine linked to a pyrimidine. See text pp. 11.

C and D are wrong, because both deoxyribose and phosphate occur in the backbone part of the molecule not in the cross link between parallel backbones. See text p. 13.

B is wrong, because ribose is not part of the DNA molecule. See text p. 24.

Question 21

C is correct.

A is wrong, because the bonds between bases are hydrogen bonds not covalent bonds. See text p. 14.

B is wrong, because peptide bonds form only between amino acids and, in any case, involve a double bond between them.

D is wrong. See text pp. 14 and 15.

Question 22

D is correct.

A and E are wrong because, while cytosine and guanine form a pair, neither pairs with adenine. So the number of cytosine molecules equals the number of guanine molecules in DNA. It is possible that, in a particular segment of DNA of a particular species or individual, the amount of cytosine and guanine might equal the amount of adenine and thymine, but this would be very exceptional. See text p. 16.

B and C are incorrect.

Question 23

E is correct.

A and D are incorrect, because neither forms a pair with a purine molecule. See text p. 11.

B and C are incorrect, as neither occurs in DNA. See text pp. 24 and 25.

Question 24

A is correct.

B is incorrect, because uracil is present in tRNA but is not complementary to cytosine. See text p. 24.

65

C is incorrect, because uracil is present in mRNA but it is not a purine and is not complementary to guanine. See text p. 24.

D is incorrect, because uracil is present in mRNA. See text p. 24.

E is incorrect, because uracil is present in ribosomal RNA but is not complementary to guanine. See text p. 24.

Question 25

1. Divide the whole sequence into groups of three. Work from left to right. (For results working from right to left, see below.) Match the RNA bases to the DNA bases in each triplet. Use the table on p. 35 to add the amino acids for the triplets.

DNA sequence→	TAG	TGC	AAA	GCT	CAG	TTC	TTA	ACG	TTG	CAT
mRNA sequence→	AUC	ACG	UUU	CGA	GUC	AAG	AAU	UGC	AAC	GUA

Amino-acid
sequence→ isoleucine phenylanaline valine asparagine asparagine
threonine arginine lysine cysteine valine

2. Working from right to left:

TAG	TGC	AAA	GCT	CAG	TTC	TTA	ACG	TTG	CAT	←—DNA sequence
AUC	ACG	UUU	CGA	GUC	AAG	AAU	UGC	AAC	GUA	←—mRNA sequence

arginine methionine ←Amino-acid
glutamic sequence
Stop acid

no more amino acids will be added after this codon has been reached. The polypeptide will leave the ribosome. See section 17.9 and section 17.10.1.

Question 26

The frequencies with which the different triplets are expected to be formed by chance associations can be predicted by combining independent probabilities through multiplication as follows:

AAA   should occur with frequency of $\left(\dfrac{8}{10}\right) \times \left(\dfrac{8}{10}\right) \times \left(\dfrac{8}{10}\right) = \dfrac{512}{1\,000} = 0.512$

AAU
AUA   should occur with frequency of
UAA

$\left(\dfrac{8}{10}\right) \times \left(\dfrac{8}{10}\right) \times \left(\dfrac{2}{10}\right) = \dfrac{128}{1\,000} = 0.128$

$\left(\dfrac{8}{10}\right) \times \left(\dfrac{2}{10}\right) \times \left(\dfrac{8}{10}\right) = \dfrac{128}{1\,000} = 0.128$

$\left(\dfrac{2}{10}\right) \times \left(\dfrac{8}{10}\right) \times \left(\dfrac{8}{10}\right) = \dfrac{128}{1\,000} = 0.128$

UUA
UAU   should occur with frequency of
AUU

$\left(\dfrac{2}{10}\right) \times \left(\dfrac{2}{10}\right) \times \left(\dfrac{8}{10}\right) = \dfrac{32}{1\,000} = 0.032$

$\left(\dfrac{2}{10}\right) \times \left(\dfrac{8}{10}\right) \times \left(\dfrac{2}{10}\right) = \dfrac{32}{1\,000} = 0.032$

$\left(\dfrac{8}{10}\right) \times \left(\dfrac{2}{10}\right) \times \left(\dfrac{2}{10}\right) = \dfrac{32}{1\,000} = 0.032$

UUU   should occur with frequency of $\left(\dfrac{2}{10}\right) \times \left(\dfrac{2}{10}\right) \times \left(\dfrac{2}{10}\right) = \dfrac{8}{1\,000} = 0.008$

Ratios of those codons in final mixture would be expected to be:

Ratio	0.008:	0.032:	0.128:	0.512
	1:	4:	16:	64
Codons	UUU	UUA UAU AUU	AAU AUA UAA	AAA

As there are four times as many isoleucines incorporated as tyrosine, and 16 times as many isoleucines as phenylanines, this suggests isoleucine is coded by 2 adenine and 1 uracil. As there are 16 times as many lysines as tyrosines, this suggests AAA is the codon for lysine. Then UUU is the codon for phenylalanine, and codons with 2 Us and 1 A code for tyrosine.

Question 27

IV—best represents the relationship.
The entry of raw materials and increase in cell size must happen together. So II is incorrect.

Increase in cell size must precede cell division so I and III must be wrong. V shows that increase in cell size follows the entry of raw material into the cell but also increase in size causes entry of raw materials to occur, and as well as this causes cell division. This does not appear to happen in living organisms.

Question 28

C is correct.

Question 29

C is correct. DNA and RNA are both substances with large molecular weights; both are similar in composition in so far that the units are base plus pyrophosphate in a 1:1 ratio.

Question 30

D is correct.

Protein synthesis involves each of the steps A, B, and C. If any one of these did not occur then no protein containing the false amino acid would be made. See text p. 32 *et seq.*

Question 31

B is correct.

A is wrong because, while ATP is involved in protein synthesis it is not involved at the stage in which codon and anticodon match, nor is it involved in making peptide bonds, by which amino acids are added to each other at the ribosomes. See text pp. 34.

C is wrong because, while polypeptides are synthesized at the ribosomes, it is mRNA that specifies the particular amino acids concerned and the sequence in which they are assembled. See text p. 31.

D is wrong because DNA polymerase is an enzyme specifically concerned with DNA synthesis, in just the same way that RNA polymerase is specifically concerned with RNA synthesis.

E is wrong because, unless there is no X present, the incorporation will take place at points determined by the interaction of tRNA for X and mRNA. See text pp. 31.

Question 32

E is correct. For polypeptide synthesis, the correct mRNA or correct substitute plus the correct amino acid plus the correct tRNA must be present.

A is wrong because poly UG does not contain the correct codon for tyrosine.

B is wrong because $tRNA_{ala}$ will not have the correct anticodon to bind with poly UAC, nor will it react with tyrosin.

C is wrong because $tRNA_{tyr}$ will not react with alanine, nor will it have the correct anticodon to bind with the codons on poly UG.

D is wrong because A, B, and C are wrong.

Question 33

B is correct.

A is irrelevant because the spiral shape does not affect the function of the molecule nor its ability to replicate.

C is incorrect, because although purines and prymidines are both found in DNA and RNA, it is possible that other molecules could function in the way that these two do.

D is incorrect. If the chemical concerned was easily changed, then the code and inheritance would be unstable.

E is incorrect because it is irrelevant.

Question 34

A is correct.

B is wrong because ribosomal RNA is part of the ribosome, not of the chromosome.

C is wrong because tRNA occurs in the cytoplasm not in the nucleus, which is the place where chromosomes occur.

D is wrong, because, although mRNA is synthesized alongside the chromosomal DNA, it leaves the nucleus and is to be found in the cytoplasm. See text pp. 24.

Question 35

C is correct.

A is incorrect, because, although the composition of a polypeptide can be discovered by means of an amino-acid analyser, primary structure requires a knowledge of the sequence of amino acids as well as composition. (See Unit 14.)

B is incorrect because the composition of some polypeptides is known.

D is wrong because enzymes—the object of study of enzymologist—are proteins.

E is wrong.

Question 36

B is correct.

A must be wrong because DNA encodes information governing the assembly of proteins.

C must be wrong because it is at the ribosomes that amino acids are assembled into proteins. See text p. 24.

D is incorrect, because without ATP or the means to produce it, all synthetic reactions in the cell would cease.

E is incorrect because the cell cannot synthesize enzymes unless it has enzymes present.

Question 37

E is correct as all the components mentioned contain nitrogen and hence ^{15}N atoms would occur in them if they were synthesized during the period in which $(^{15}N H_4)_2SO_4$ was the source of nitrogen for *E. coli*.

Question 38

D is correct as both ribosomes and mitochondria contain proteins. C is wrong because mRNA is not a protein.

Question 39

E is correct.

A is wrong because it goes much further than the evidence warrants. The DNA added may be doing other jobs in the cells to which it is added, besides replacing or acting together with the existing DNA in the cell.

B is wrong because the molecular weight of any substance does not determine its chemical behaviour. The similarity of molecular weights may be fortuitous.

C is wrong, because although the functions of the DNA made in the test-tube and the naturally occurring DNA are similar in one respect, they may not be in others including their chemical composition.

D is wrong because A, B, and C are wrong.

Question 40

A is correct.

B, C, D, and E are all examples of structure and function being related. See Unit 14 and, for mitochondria and chloroplasts, *The Microstructure of Cells*, Chapters 6 and 13.

The spherical shape of the nucleus is irrelevant to its function of DNA storage—presumably any other shape would be possible and indeed not all nuclei are spherical.

Answers to Questions in the Main Text

Section 17.2.2 Structured Exercise

Question 1

System A   No. If two strands were made while the parent cell was in a ^{15}N enriched medium, and two more strands were made while the parent cell was in the medium without ^{15}N, then the strands made first would have a higher density than the later strands.

System B   Yes. In this case old (^{15}N containing) and new (lacking ^{15}N) strands are intermixed, so that the weight of all four strands would be similar.

System C   No. The argument here is the same as that given in A above.

Question 2

In system A, a 1:1 ratio of heavy to light strands.

In system B, all strands with the same density.

In system C, as in system A, a 1:1 ratio of heavy to light strands.

If you are uncertain about these answers check back to the diagram on p. 20.

Question 3

The results show an equal amount of heavy and light DNA after one generation in the medium lacking ^{15}N. *All* the DNA in cells put into this medium contained ^{15}N and must therefore have been heavy. This is a consequence of growing the bacteria for many generations in ^{15}N enriched medium. The light DNA can only, therefore, have been made after the parental cells had been put into the medium lacking ^{15}N. So, as equal amounts of heavy and light DNA are present after the cells have divided once, then system B must be incorrect. If system B were correct, then all the DNA should have had the same density. System A and system C both predict the result correctly and so either might be correct.

Question 4

No. Either system A or system C could be correct. The argument is given above.

Question 5

That system B is not the way in which DNA replicates. We cannot at this stage eliminate either system A or system C, as both stand up to experimental tests so far.

System A Generation 1

←— Lighter   Heavier —→

Grown in ^{15}N medium

Grown in medium lacking ^{15}N

System C Generation 1

←— Lighter   Heavier —→

On the basis of system A, a pair of heavier and a pair of lighter strands would be expected. So two bands would appear in the result. On the basis of system C, although two light and two heavy strands would be expected, these would pair up one heavy with one light chain—so the two pairs would be the same density. This would produce only one band, not two. The density of the band would be half-way between that of the light DNA and the heavy DNA.

Question 7

Clearly system C—because all duplex (= double-stranded) DNA consists of one heavy and one light strand, and thus has an intermediate density, whereas in system A all the double-stranded DNA would be either heavier or lighter.

Question 8

There are four features of the experimental results which seem to stand out:
  (1) disappearance of heavy DNA after transfer to new growth medium;
  (2) appearance of light DNA after two generations, i.e., one division cycle;
  (3) persistence of DNA of intermediate weight during whole investigation;
  (4) increase in amount of light DNA relative to amount of intermediate DNA.

These four features can all be explained if DNA replicates in a semi-conservative manner. A duplex of one heavy and one light strand produces two daughter duplexes; one of one heavy and one light strand together, and the other of two light strands together. So the original amount of heavy (^{15}N) DNA remains constant, while the total amount of light (^{14}N) DNA increases at each replication. The process is shown diagrammatically below:

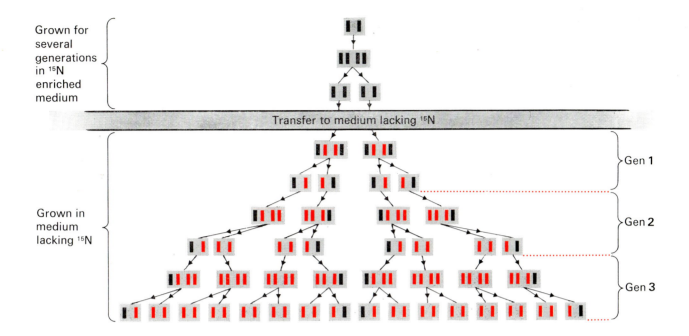

Grown for several generations in ¹⁵N enriched medium

Transfer to medium lacking ¹⁵N

Grown in medium lacking ¹⁵N

Gen 1

Gen 2

Gen 3

Section 17.7 Questions 9–12

Question 9

If (a) is correct, then protein of progeny should be type B protein.
If (b) is correct, then protein of progeny should be type B protein.

Question 10

Presumably a mixture of both type A and type B protein, or perhaps a protein that was neither type A or type B. But in this circumstance it seems certain that the protein made would not be *all* either *A* or *B* type.

Question 11

It can't be, unless the determinations are subject to a large experimental error. Such an explanation might save the hypothesis in the case of histidine, but becomes less credible for the methionine discrepancy. The prediction made in answer to Question 2 fails therefore.

Question 12

No—none at all.

Section 17.11 Questions 13–25

Question 13

One crucial question is, 'was the root actually growing at the time when it was killed?' Many other questions could be important and it is *not* possible to make a complete listing. This is because it is not possible to say what all the information relevant to any situation is. However, it would be useful to have information about how the plants were grown, how old they were, the time of day and the time of year when the root was killed, how the root was killed, how long did it take to die, what happened to the cells during the death process and many other questions. The questions about killing are particularly important, because changes occur during the killing of cells which alter their appearance and produce artifacts. Artifacts were mentioned in Unit 14.

Question 14

If, at the time of their death, the cells look similar, it suggests they were all at a similar stage in the cell multiplication process. They were in step or synchronized. In the running analogy, the running people would all have looked similar because they were in step.

Question 15

No. You would need a sequence of diagrams or photos taken at different times. You would need something like a movie film, either for a multiplying cell or the running people.

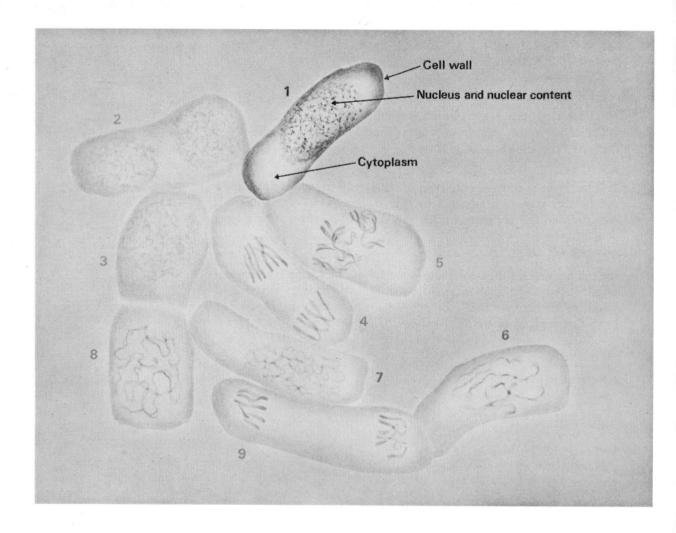

Question 16 and Question 17

Cells 2, 3, 12, 13, 14 are all very similar in appearance to Cell 1; so, as in Cell 1, it is possible that a nuclear membrane is present in each case.

The other cells are much less like Cell 1 in appearance and have the nuclear material more spread out inside the cell; so a nuclear membrane is less likely to be present.

In fact, in multiplying cells at the stages seen in the photograph in Cells 4, 5, 6, 7, 8, 9, 10, 11, and 15, a nuclear membrane is definitely not present.

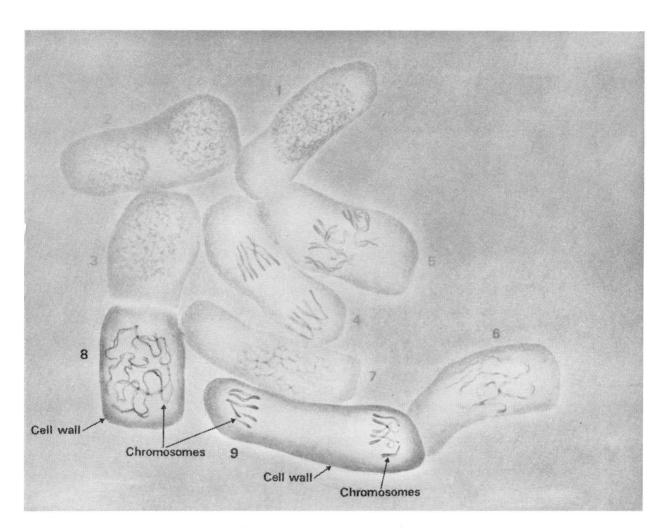

Question 18

In Cell 8 the chromosomes are:	In Cell 9 the chromosomes are:
Longer	Shorter
Thinner	Fatter
Double Threads	Single Threads

Question 19

In Cell 9, the short thick chromosomes are arranged in two groups in the cell, while in Cell 8 the long thin chromosomes are in a single group in the middle of the cell.

Group A cells—Cells 1, 3, 13, and 14

Group B cells—Cells 2 and 12

Group C cells—Cells 4, 9, and 11

> This category depends essentially on the interpretation of single threads down their whole length.

The chromosomes in some cells appear to be single rather than double threads for *part* of their length—so they are excluded from this class.

Group D cells—Cells 7, 8, 10, and 15

Group E cells—Cells 5 and 6

Considering both these classes together the first characteristic is clear cut: only part of each chromosome need be double for that cell to qualify. The relative lengths and thicknesses of the chromosomes are more difficult to decide in some cases. Cell 6, for example, might with justification be classified in the same group with Cell 8. In fact, for present purposes, it wouldn't greatly matter if it was. Remember that dividing the multiplication processes into discrete stages is an arbitrary proceeding and is a means to an end rather than an end in itself.

Question 20

Group A is correct

Question 21

Group B is correct.

Remember that the process being examined is a multiplication process—so reasonably there ought to be more cells at the end, after cell multiplication, than at the start. So Group B—a single cell with two nuclei—is the most likely type to find near the end of multiplication.

Following this, a cell wall grows across the cell separating two new nuclei and completing the division. This is not seen in any of the cells in the drawing.

It is not possible, with the data available in the drawing, to decide absolutely if any of the cells in Group A are really cells which have just been finally separated from each other or if they are cells in the very early stages of a new multiplication cycle. For our purposes it does not really matter, just because multiplication is cyclic.

Question 22

Start (A) → D → E → C → Finish (B) is the correct sequence. The whole process is shown in the television programme of this Unit.

Question 23

In stage C, the chromosomes appear to be short, thick single threads. In neither stage B nor stage A are chromosomes visible. But in stage D, the long, thin chromosomes can be seen to be double down part (greater or smaller) of their length. If we assume that there is continuity of structure of the chromosomes, even during the stages when the chromosomes do not react with the stain used, then obviously the chromosomes must get longer and thinner, and either double (replicate) or divide along their length. The information you already have concerning the manner in which DNA replicates (section 17.2.1) rules out the possibility that:

(a) getting longer is due to DNA being added to the ends of chromosomes;

(b) there is simple lengthwise division of the chromosomes.

Changes in length and thickness of the chromosomes during cell multiplication are largely due to the chromosomes becoming tightly twisted and hence shorter and thicker, or becoming untwisted again and so longer but thinner. These movements are not as important as the change from single to double threads. One of the essential features of mitosis is the replication of the genetic material that precedes cell multiplication. When this is complete, the chromosomes look double.

Question 24

As each chromosome contributes one of its two parts to the daughter cells, and as these two parts are derived by semi-conservative replication from pre-existing chromosomes, then it follows that the daughters inherit similar characteristics to each other and to their parent cell. This exact distribution of chromosomal material is the second important feature of this type of cell multiplication. Variety of inherited characteristics is a consequence of a second type of cell multiplication called meiosis, which will be considered in Unit 19.

No very satisfactory description of the physical relationship between chromosomes and DNA molecules has yet been written. While it is certain that chromosomes are made up of DNA and other materials (largely proteins), we do not know the details of the chromosome structure at the molecular level.

Question 25

Provided each daughter cell is the product of a mitotic multiplication, then it follows from the argument used in the answer to Question 24 that all daughter cells will resemble each other and the zygote from which they originated.

ACKNOWLEDGEMENTS

Grateful acknowledgement is made to the following sources for material used in this Unit:

TEXT

McGraw-Hill Book Company for W. D. Stansfield, problem 9.3 in *Theory and Problems of Genetics* on which Question 26 is based.

ILLUSTRATION

The American Association for the Advancement of Science for *Fig. 8* after F. C. Steward in *Science*, Vol. 143, 3. January, 1964.

Cover Plate

An anonymous seventeenth-century engraving, Creation of Homunculus. *The art of alchemy and a prayer for divine blessing combine in the creation of an artificial man.*

Page from the Schweizerisches pharmaziehistorisches Museum, Basel.

Notes